SUCCESS WITH CHICKENS

THE 'WHAT, WHERE AND WHY' OF TROUBLE-FREE CHICKEN-KEEPING

J.C. JEREMY HOBSON

All photographs by Robert (Rupert) Stephenson
unless otherwise credited

First published in the UK in 2012
by Quiller, an imprint of Quiller Publishing Ltd

British Library Cataloguing-in-Publication Data
A catalogue record for this book
is available from the British Library

ISBN 978 1 84689 093 2

Printed in China
Book design by Sharyn Troughton

Quiller
An imprint of Quiller Publishing Ltd
Wykey House, Wykey, Shrewsbury, SY4 1JA
Tel: 01939 261616 Fax: 01939 261606
E-mail: info@quillerbooks.com
Website: www.countrybooksdirect.com

CONTENTS

ACKNOWLEDGEMENTS

If you are enthusiastic about any subject then writing about it is the easy bit. If your subject happens to include livestock of any kind – all of whom have a mind of their own, then photographing the subject becomes nigh on impossible – unless, of course, you are possessed with the same degree of knowledge and patience as Rupert Stephenson, a professional poultry photographer who has supplied virtually all of the illustrations for this book. Christened Robert, but known to all in the poultry world as Rupert, he has, as always (I've been fortunate enough to have been able to persuade him to provide photos for several previously published books and many of my articles), come up trumps and didn't seem at all fazed by the rather daunting list of requirements I sent him. My first and most sincere thanks must, therefore, go to him; not only his most valuable contribution to this title, but also his friendship – one day we might actually meet rather than communicate via emails!

There are a few photographs from other sources and I am most grateful to David Bland; Elliot Hobson; *Flyte so Fancy Ltd*, Dorchester, Dorset (who were most generous in giving me access to several photos from which to choose); *Green Frog Designs*, Templecombe, Somerset;

Crofting Supplies, Caithness, Scotland and *Osprey Plastics (BEC UK)*, Church Stretton, Shropshire.

Any chicken knowledge I do possess didn't just spring into my head as a lightning flash. Over forty plus years, I have gleaned much from a vast number of people, but those who remain constantly in my mind and who are sadly no longer with us include my maternal grandfather; A. E. Walker; Jim Ellis and George Lodge. More recently (and thankfully they are all still very much alive and kicking!) my knowledge has been much enhanced by the likes of David Bland, Tony Weatherill, Peter Delaney and Charlie Sullivan. Specifically appertaining to *Success with Chickens*, I would like to thank Tim Nelson, headmaster of Staverton Church of England Primary School, Daventry, Northamptonshire, for allowing me to mention his quite innovative scheme (page 27) and to quote some of his words which originally appeared in *Your Chickens* magazine – appertaining to which, grateful thanks are also due to Simon McEwan, *Your Chickens* Content Editor, for agreeing to their use and to my making reference to other articles which have appeared in his excellent publication. Also, a big 'thank you' to all those who recounted their chicken-keeping exploits as they appear in Chapter 2.

In the past and in order to find suitable magazine subject matter, I have occasionally stolen ideas from Valerie Porter's book *Domestic and Ornamental Fowl* (Pelham, 1989). If she looks carefully through these pages she might find I've done so again and I know she'll have no objection – I must, therefore, thank her not only for her leniency, but more sincerely, for a friendship which has lasted for almost thirty years!

As regards any of the other books mentioned or quoted from in the text (for anything more than a few words), I have checked and they are, to the best of my knowledge, either out of copyright or their original publishers cannot be found. If I am erroneous in my thinking, or have been a little lax in my research, I most sincerely apologise and ask that anyone offended contacts the publishers in order that the situation can be rectified in future reprints.

1
WHICH CAME FIRST...?

Marco Polo described a breed of chicken that can only have been the ancestors of the modern Silkie

It is, I know, something of a cliché to ask the question but, which exactly *did* come first, the chicken or the egg? Without going anywhere near Darwin's theory of evolution, one has to suppose that the chicken was needed to lay the egg; but from where did the chicken come – out of an egg is the obvious answer! Depending on how many millennia we want to go back, some authorities have it that birds of all kinds developed from reptiles and their feathers from scales, but as modern-day chicken-keepers, do we need to know even this tiny piece of information? Probably not. What is, however, generally accepted is that the ancestors of today's chickens were almost certainly the red Jungle Fowl of south-east Asia which were reputedly first domesticated by the people of the area some three thousand years BC. In more recent times (these things are all relative!) there are records of domestic chickens being kept by the Egyptians, Greeks and Romans. How chickens eventually came to populate the whole world is undoubtedly a result of exploration and trading; for example, Marco Polo mentioned 'fur feathered' animals, which were undoubtedly the ancestors of the modern-day Silkie.

DEFINING 'SPECIES' AND 'BREEDS'

Before moving on to discuss the make-up of chickens, it might be as well to make clear the distinction between species and breeds. A species will, in the wild, only ever mate naturally with its own kind, partly because it recognises the appearance of another in the same species as being similar to its own, partly because of circumstance (habits and breeding seasons must overlap, for instance) and partly because the offspring of two different species, should such an unlikely meeting ever take place, would most likely be non-viable from the moment of conception or infertile and unable to reproduce due to a mismatch of chromosomes.

Warren hybrids – good egg layers and easy to keep, but sadly not much to look at when compared with some of the pure-breeds!

All chickens are the same species and can interbreed quite naturally, but an Ancona is a different breed to a Yokohama. Basically then, a pure-breed is perhaps easiest to define as being a true genetic breed, i.e. when male and female of the same breed are mated together, they are guaranteed to reproduce virtually identical offspring. To define it further, one could say that a particular breed is a group of birds that have been produced over generations and nowadays possess inherited characteristics such as shape, colour and comb formation, which all help to distinguish it in some way from other birds within the same species. Just where this leaves hybrids and cross-breeds is a matter which is discussed in Chapter 4.

THE MAKE-UP OF A CHICKEN

Evolution over millennia has developed a very successful end product. The chicken's beak is designed for pulling at vegetation and pecking around for grubs, on-ground seeds and grain; it is, therefore, relatively thick-set and fairly robust – more like that of a berry-picking finch than a probing long-beaked insect-seeking woodcock. Technically, the beak is split into what are known as the upper and lower mandible, and the mouth part, of course, contains no teeth – instead food is passed directly to the crop and subsequently into the gizzard whose rough interior will, with the aid of grit, grind down the food into a sort of paste, the nutrients from which can then be used by the bird's system.

More or less all-round vision is vital to what was once a wild bird, which needed to be constantly on the look-out for potential predators as well as food, and the eyes are set into the side of the head rather than at the front where vision would be limited. Tucked away behind the eye, and just above the wattles, are the ears – not always immediately obvious unless you know which feathers to brush aside in order to look.

Despite what people may think chickens are quite brainy birds and, according to Siobhan Abeyesinghe of the Silsoe Research Institute, 'do not just live in the present, but can anticipate the future and

Chickens were not the first domesticated poultry

Although, as I would be the first to admit, it has nothing at all to do with the origin of the chickens in your back garden, I nevertheless thought that the following might prove of interest when it comes to comparing the domestication of chickens with other farmyard fowl.

There is evidence that geese were domesticated as early as five thousand years BC (undoubtedly much earlier than chickens), as part of the process of socialisation that saw humankind move away from nomadic herding to settled farming. The remains of domestic web-footed fowl discovered among the grave goods in certain prehistoric burial sites in central northern Europe date from the Neolithic period and while the people of south-east Asia were just getting into their stride as chicken-keepers, in Egypt, wall paintings dating as early as the 5th Dynasty show the practice of force-feeding geese. This suggests that as long ago as two thousand eight hundred BC there were those who knew how to appreciate the fine qualities of *foie gras*!

The head of a Spanish cock bird clearly showing the beak, comb, eye and wattles

demonstrate self-control, something previously attributed only to humans and other primates'. Her findings, as a result of research carried out in 2005, suggest that chickens are 'intelligent, forward thinking creatures and may therefore worry and be capable of expectations, leading in turn to feelings of thwarting, frustration and pre-emptive anxiety'. Research undertaken at Bristol University in 2011 has also shown that chickens are able to empathise with one another – a hen, for example, has been proven to feel stress and her heartbeat increases when one of her chicks is even slightly distressed. Although there is still more work to be done, it seems likely that unrelated birds will feel the same empathy – a point to bear in mind when catching a chicken for whatever reason.

Sometimes known as 'head furnishings' or 'furniture', the most distinctive feature of a chicken is probably its comb and wattles. The main reason for their existence is to act as a cooling system because, like a dog (which uses its panting tongue to lose heat), birds cannot sweat, so the chicken cools itself by circulating blood through the comb and wattles, thereby dissipating body heat. Bearing in mind that breeds such as the Ancona, Minorca and Leghorn originated in Mediterranean countries, which are hot for a good portion of the year, it is logical that their combs should be, in general, large – and they usually stand proud or flop away from the head, giving the largest possible expanse of area through which heat can be lost. What makes less sense, however, is the fact that some breeds such as the Malay come from even hotter places and yet only have a walnut comb and virtually no wattles to speak of – they are relatively sparsely feathered though, which may help. Comb types are many and varied, but include what are known as single, horn, rose and pea – as will become more apparent in Chapter 4.

Feathers provide both insulation and waterproofing. The colour of a feather to some extent affects its resilience; for example, black feathers are thought to be more resistant to wear and tear than white ones because they contain more pigmentation. The main pigments are melanin (manufactured in the bird's body) and carotenoids (which are absorbed from foods, especially greenstuffs and roots). Not all birds have the same amount of feathers – some breeds have them all the way down their legs and may also have feather beards and crests for protection in winter weather (most of these also have almost non-existent combs, so the combination of small combs, muffs, beards and crests must be Nature's way of preventing heat loss). The need to fly has become redundant in many breeds, but the wing feathers are split into what are known as 'primary' and 'secondary' (the large ones that can easily be seen along the outer edge) which developed in such a way as to make flight possible. Some feathers are, of course, purely ornamental and designed to attract the opposite sex: the sickle feathers in the tail, the saddle, wing shoulders and the neck of the cock bird, for instance. Neck hackles, however, have not only a sexual purpose and they are also intended to make the cock look far more aggressive in response to any possible threat to his harem. It is not only the obvious contenders that might face a sparring cock bird with his hackles raised and, despite generations of domesticity, a cockerel in India will adopt exactly the same position upon encountering a snake. Anyone who has kept chickens for many years (especially the often tenacious bantam) will, no doubt, have suffered at some time or another as a result of a

The cockerel waltz

Some of the more aggressive males will not only raise their hackles, but also drop and extend both wings and puff out all their body feathers to give their harem (or other cocks) the impression of greater size. In what is sometimes known as the 'cockerel waltz', a cock bird will drop the feathers of just one wing and scuttle round an individual hen in a display of dominance. More often than not, the female will either move away or more likely, squat in submission, making it much easier for the cockerel to mount and mate with her.

Most breeds have four toes, but some, such as the Dorking and Faverolles, have five

particular male bird who insists on flying at the back of their legs, hackles set in a ruff formation and spurs ready to do the maximum amount of damage!

The chicken's legs are scaly and, as has been pointed out, may have a covering of feathers. In fact, some breeds have not only feathered legs, but feathered feet as well – which does not always make them an ideal bird for the inexperienced chicken-keeper (as will be explained later). Cock birds possess spurs (which, as the bird grows older, may cause problems when mating if they are not trimmed occasionally) and on some hens, little nodules can be seen at the back of the legs. The toes are strong and perfect for scratching about in order to find food or to create a shallow basin in which they can dust-bathe and help rid themselves of parasites. Most breeds have four toes (one being at the back) but some, such as the Dorking or Faverolles, quite famously have five.

THE MAKE-UP OF AN EGG

There are so many references to eggs in common usage nowadays and yet so many originate from many years ago – and not necessarily from the UK. 'Don't put all your eggs in one basket', was not, as might be imagined, one of a prudent Victorian grandmother's old saws, but was first mentioned as an Italian proverb of 1662. For other proverbs concerning eggs, look no further than the practicalities of cooking: 'You cannot make an omelette without breaking eggs' is a simple and most logical statement.

Eggs, therefore, are all things to all people: the back garden chicken-keeper will expect great things of theirs, irrespective of whether they are intended for eating purposes, or to be collected and carefully saved for hatching. Eggs, to a lesser or greater degree, provide all the essential amino acids and minerals required by both the developing chick (in the

Egg-lovers will no doubt be pleased to learn that research has proven that their cholesterol content of eggs is nowadays much lower than a decade ago!

case of a fertile hatching egg) and the human body (when used for cooking). They are an important source of vitamins A, B, and D for humans as well as supplying a complete, high-quality protein source. Some people avoid eating eggs in the belief that they contain high cholesterol levels; however, egg-lovers will no doubt be pleased to know that the cholesterol content is much lower when compared with ten years ago – at least according to a study carried out in early 2011. The reason eggs have apparently become better for you over the past decade is that hens are no longer fed bone meal, which was banned in the 1990s following the BSE crisis – so now you know.

Eggs over the roof-tops!

In the past, a yolkless egg has been the subject of much superstition and been given many names such as 'wind' eggs or even 'cock' eggs – the latter because the ancients thought that, as the shell was yolkless, it couldn't have been laid by a hen and must, therefore, have been laid by the male of the species (no doubt there was some logic to their argument, but I fail to see it). Worse was the belief that these 'cock' eggs would, if set, hatch out a huge, evil, snake-like creature that was capable of causing death by just looking at someone. The only sure way of avoiding such a fate was to throw the egg over the roof of the family residence so that it broke into tiny pieces on the other side. The thrower had to take great care that it didn't hit the roof; fortunately, that was less likely then to have required the skills of a rugby player as the dwellings of the time would have been low, single-storey affairs.

Structurally, an egg is quite a simple piece of kit. Protected by a porous shell (made up of calcite – a crystalline form of calcium carbonate), its contents include not much more than the albumen (or white), made up of water and protein, and the yolk, all of which is protected by 'shock absorbers' known as chalazae. This simple form, which is nevertheless capable of producing the miracle of life, has been a source of wonder and a symbol of rejuvenation for thousands of years.

Occasionally, and most likely at either the beginning or the end of a period of laying, a hen may produce an egg with no yolk at all or, conversely, one with a double yolk. An egg with no yolk is quite likely to be smaller than average, whereas a 'double-yolker' will, more often than not, be larger than the norm. The minimum weight that used to be aimed for when it was common practice to trap nesting birds in order to keep a record of their egg production was 55g (2oz), but the newcomer to the trap record system was always urged not to be too severe on young pullets because, until a bird is fully matured, the eggs she lays are often a little smaller and lighter than the ones she produces once she has reached adulthood.

Eggshell colour

Egg colour depends mainly on the breed. As a general rule (but not always the case – *see* page 152), the lighter Mediterranean breeds all lay whitish eggs, those heading towards being dual-purpose lay creamy, tinted, or light brown eggs, and some of the heavy types lay brown, occasionally bordering on deep chocolate eggs – the most notable of these being Marans and Welsummers.

What is it that makes a tray of dark brown eggs visually more attractive than a tray of white ones? Provided that the hens that lay the brown eggs have been given exactly the same environment and feeding regime as the ones that laid the white ones, they do (and it has been scientifically proven) all taste the same. Nevertheless, position a basket of white eggs at the front of a stall at a farmers' market and one of brown eggs at the back and you will find that, despite being less prominent, the brown eggs sell

The colours of eggs vary greatly from breed to breed

more rapidly than the white. Interestingly, were you to conduct the same experiment in America, you would most likely find that the white ones were favoured – something to do with white indicating cleanliness and purity in the mind of the average American consumer, or so I'm told.

Blue eggs are sometimes laid. The Araucana Club of America *(see also* page 75*)* has carried out a great deal of research on the subject and, until relatively recently, thought as I did, that their particular breed was the only blue egg layer in existence (with the possible exception of the blueish-coloured eggs laid by Cream Legbars, a hybrid). Their research has, however, discovered another source from the Jiangxi Province of

China. Known as the Dongxiang, the breed, or something approximating to it, has apparently been around for several thousand years. It is black fleshed rather like the Silkie and is slightly smaller, but nevertheless more heavily built than the Araucana. Dr Ning Yang is a poultry and genetics expert responsible for the well-being of the gene pool of all seventy-eight species of native Chinese chicken breeds and he says that, as far as he can ascertain, none of these birds has ever left China and, in addition, there is no history of an Araucana-type bird ever being introduced to China. It seems likely, therefore, that these are two completely separate breeds.

It is, after all, the contents of the shell that we eat!

However, unless you are committed to showing eggs (*see* Chapter 8), provided that you are doing all that is possible to ensure that good sized eggs are being hatched, that their colours are as defined by the breed standard and they come from hens that are producing the number of eggs expected from the particular breed, the depth of shell colour is not so very important – it is, after all, the contents of the shell that we eat (or hatch) and not the shell itself.

Egg yolks

The yolk is made up of vitamins, minerals, protein, water and fats, and is intended for use as a food store by the chick on hatching (it absorbs the yolk prior to chipping through the shell). The colour of the yolk is directly associated with the hen's diet. F. E. Wilson, writing in 1903, claimed that: 'it should not be forgotten that the colour and flavour of eggs depends very much on the type of food used. Hens fed largely on wheat frequently lay – especially if kept in confinement – eggs in which the yolk is of the palest yellow; whereas hens at liberty, getting plenty of worms and insect food, lay eggs with rich dark-coloured yolks…it will therefore be readily seen that care needs to be exercised if well and finely-flavoured eggs are desired.'

Provided that a bird is given ample opportunity to scratch about and find natural food for itself (and some seeds and insects on which it feeds are not immediately obvious to the human eye) and is fed a balanced ration bought from a reputable feed manufacturer, there should be no problems with the quality or colour of the eggs. On occasion, when showing eggs for example, the addition of a small amount of maize to the diet will help to give the yolk an almost golden colour but great care must be taken when feeding this particular cereal as too much will result in vast amounts of fat being deposited in the bird's body – which will do no good at all.

Bantam eggs are thought to contain a much larger yolk-to-white ratio than is normally found in the eggs of large fowl. When it comes to the question of whether or not bantams lay as many eggs as their large fowl counterparts, some breeds do in fact perform remarkably well. Naturally they have not been bred for egg production in the same way as some of the commercial chicken types, and of course what eggs they do lay are proportionally smaller. Nevertheless, many bantam breeds can match their larger cousins when it comes to the number of eggs laid. They are therefore worthy of consideration if space is limited and you simply require a few birds that will enhance your garden, create interest and give pleasure, plus a reasonably regular supply of eggs.

BANTAMS – WHAT ARE THEY?

The generally accepted definition of bantams distinguishes 'true' bantams (for which there is no large breed counterpart) from 'miniaturised' or 'diminutive' fowl which are anywhere between one quarter and one fifth of the size of their large breed equivalent. The miniaturising of large breeds probably began in the late sixteenth century, but it was not until the Victorian era that bantams became really popular and breed standards began to be set; these standards may have changed slightly in the intervening years but I think that most breeders and enthusiasts back then would, by and large, recognise the majority of those we see today.

As most (but certainly not all) large fowl breeds have been miniaturised, you should be able to find a bantam version of the breed that particularly fascinates you. In addition (although I might be biased as they are my first love), they are, I think, possibly just a little more characterful than some of the large fowl breeds and will most certainly keep you amused with their antics as they follow you enquiringly around the garden. W. H. Silk, in his treatise *Bantams and Miniature Fowl* (published in 1951) was, I think, waxing lyrical when he penned his personal definition of bantams.

Bantams are perfect in a garden environment

> WHAT IS A BANTAM? The old-timer will glibly tell you that it is a genus of miniature fowl which originated at Bantam in Java. He won't know whether he's right or not; and it is much more likely in truth that the Japanese were mainly responsible for originating them.
>
> The die-hard exhibition man will describe it as a species of small fowl…oblivious of the fact that bantams originally had no counterparts in big poultry.
>
> The farmer will say it's a child's pet. Many breeders of large fowl will call it a pest that should be discouraged; but that is because nowadays bantams have usurped the popularity that large breeds formerly held at shows.
>
> All of them, of course, are wrong to a greater or lesser degree. The bantam is no mystery, neither is it a commercial money-spinner. On the other hand, it isn't (or shouldn't be) a freakish unproductive dwarf.
>
> It is capable of production as well as reproduction, and is, in fact, a fascinating hobby, and a bundle of charm. It won't read your thoughts like a dog, or purr against your legs like a tame cat…but within its sphere those with limited space and little cash will find it supreme as a combined source of relaxation, home profit and amusement.

SOME 'CELEBRITY' CHICKEN-KEEPERS

Keeping chickens is, or should be, a restful pastime. If work commitments dictate that you are away for much of the time and there is no one in the family likely to share your hobby, or you are of an impetuous, loud and blustery nature whereby dogs cringe when they see you and cats just slope off somewhere quiet, then perhaps you are not best suited to the keeping of any kind of livestock at all. Thankfully, the majority of people are, and even the busiest and most famous of people find some sort of calm, refuge and amusement down in the chicken run.

Apparently, as a child, the Irish novelist Cathy Kelly (appropriately enough, a writer of what is now often referred to as 'chick-lit') spent a great deal of time with her grandmother who kept chickens: '...it was my job to watch the young chickens in the morning because they'd rush off to lay their eggs in hidey holes and I was supposed to find them. I loved those hens. I taught them to jump for blackberries and gooseberries. You wouldn't think hens could do that, would you?'

Surprisingly, given her obvious interest in cooking and obsession with the countryside, Clarissa Dickson Wright has yet to succumb to the charms of the hobby, but, as she remarked in her book *Rifling Through My Drawers* (Hodder & Stoughton, 2009), 'one of the things I want to do before I die is keep chickens...'. Pam Ayres, Amanda Holden, Billie Piper, Sadie Frost, Kirsty Gallagher, Julia Kendell, Anthony Worrall Thompson and Jamie Oliver are all well-known back garden chicken-keepers, as is TV presenter Philippa Forrester who has several rescue hens from the British Hen Welfare Trust as well as a 'crazy Silkie bantam who looks like a white feather duster on legs'. Asked why she chose the rescue birds, Philippa says, 'When they first came they had never felt the sunlight on their backs or had the room to scratch or spread their wings and every time I watch them now scratching in the grass, pecking up bugs and having a sunbathe, every morning when I let them out, it gives me a deep satisfaction that I have transformed their lives.'

2

WHY KEEP CHICKENS?

A breed such as the Sussex is an attractive proposition to anyone who wants eggs, a garden pet, or to exhibit at show standard level

While some people are pleased to have transformed the lives of their rescue hens many more claim that chicken-keeping has transformed their lives. One high-profile owner who, understandably, asked to remain anonymous, states that she 'was in a dark place until a friend turned up one morning with a trio of hens, a coop and run and a bag of food – together with the comment "if I find you've not been looking after these, there will be big trouble". Having to be responsible for something other than myself was a good incentive to get up in a morning without first going in search of the gin bottle.'

Less dramatically, countless back garden chicken-keepers (it is estimated that there are at least half a million such people in the UK) have found a great deal of fun, achievement and contentment from their hobby. Whilst there have been times (during World War II for example) when they have been kept by all who could as a valuable source of eggs and meat, it seems that never before have so many been kept simply for the pleasure they give. The hobby appears to be a growing trend on both sides of the Atlantic, and even in Europe – where it has always been common amongst rural dwellers – there is an ever-increasing interest in pure-breeds, rare breeds and exhibitions.

Perhaps rather than ask 'why keep chickens?' it might be of more value to ask 'what do you expect from your chickens?' because from the answers will come most of the reasons why you should keep chickens! It could be the simple enjoyment of having a few attractive birds in the garden and being able to enjoy a home-produced egg or two for breakfast or when cake-making. Parents worried about their children's seeming addiction to electronic games and gadgets might think that getting their offspring interested in a new pastime will make them fitter and more interested in life outside the sitting room (and so it will – chicken-keeping is a brilliant way to teach youngsters to be responsible). Others, inspired no doubt by the welfare issues concerning commercial

Not so 'common or garden' bantams enjoying the garden! These are, in fact, a prize-winning trio of Dutch lemon porcelain

poultry production, profiled in recent years by the likes of Hugh Fearnley-Whittingstall and Jamie Oliver, wish to give ex-battery or otherwise intensively housed chickens an extended period of life, whilst some are fascinated by the breeding side of things and want to perpetuate the genes of a rare breed. Yet more people hope to combine several of these aspects and show their birds at local, county and national level.

How and why?

The reasons why people want to keep chickens are, as we have just seen, many and varied – although I suspect not many can emulate one lady who told me that she got started as a result of being given a trio of Buff Orpington large fowl by the late Queen Mother! More ordinary, but no less interesting, are the comments of the following hobbyists and fanciers. Graham Langston, now a free-range poultry farmer in Yorkshire, still keeps a few rare breeds and remembers how he got started: 'Mum showed goats and I used to go with her to the agricultural shows, where I always made a bee-line for the poultry tent and stood for hours gazing at the Belgium bantams. Mum got to know one of the breeders and bought me a trio as a thirteenth birthday present – she also ended up marrying the breeder, but that's another story entirely!'

Not all parents were so encouraging. 'I'd been pestering for some chickens ever since getting involved with the Fur and Feather Club at school', says Peter Finchley. 'My parents held out as long as they could but then there were some Light Sussex/Rhode Island Red youngsters that had been hatched and reared in the school incubator which were looking for a temporary home during the summer holidays. Under duress, my parents agreed to let me bring them home, where they soon settled in a partitioned part of the garden shed. They never went back; I fattened and killed the three cockerels and sold the eggs from the pullets to neighbours and family friends.' Helen Munro started keeping chickens by accident. 'A friend of my father who kept and showed Old English Game bantams died suddenly and to help his widow out, my

The most popular bird in the world

Charles Trevisick, who was in his time a farmer, zoo owner, author and television personality amongst other things, told of when he was on a Q & A panel and was asked the question, 'Which is the most common bird in the world?' His mind, he said went blank, but then 'The penny dropped. "The domestic fowl," I replied, and took a sip of water knowing I was right. Yes, your common or garden chicken is the most popular bird in the world, and with good reason, for whether your back garden is in Glasgow or Seattle…you need only set her down with a patch of earth to scratch, some food to eat and a corner to nest in and she will do you proud.'

As recounted in his book, *Keep Your Own Livestock*, Stanley Paul & Co. Ltd, 1978 dad offered to take his five pens of breeding stock off her hands. Dad then had to change his job, which meant he was away from home for days on end and so I was given the job of looking after the birds. Eventually I took over completely and bought a pen of Silkies to use as broodies. Soon I was hatching and showing both the OEG *and* the Silkies!'

As many people have found, an original 'flock' of two or three chickens has a habit of growing in numbers! *(Photo courtesy of Elliot Hobson)*

Many of today's chicken-keepers started with one particular breed and then, for a variety of reasons, moved on to another. 'My first chickens were Brahmas', recalls Anthony Simpson. 'I used to show them but unfortunately at the time had to enter under the Any Other Variety (AOV) classes due to there being never enough entries to warrant the breed being given a class of its own. Although I occasionally won, I didn't think it a very fair competition due to the judges not being able to compare like-for-like so I changed and have been keeping large fowl Welsummers for over twenty years. Ironically, I've never bothered to show them!' Margaret Brant kept semi free-range hybrids until her husband's work forced a move to an area where houses were more expensive but nevertheless stood in smaller surroundings, thus forcing her to sell her hybrids and instead keep five Ancona bantams in a combined house and run that can be moved around the garden.

CHICKENS AS ALLIES TO GARDENERS

Provided that you do not leave a small house and run on the same patch of lawn for long periods (try to move it every day if at all possible), the scratching of your birds will do much to remove unwanted moss, which can then be raked up and added to the compost bin where it will rot down and eventually create nourishment for the vegetable patch. Interior designer and television presenter Julia Kendell claims that 'there's something very special about having chickens around. I love to grow vegetables and spend a lot of time in my garden. I cannot imagine my garden without them!'

Vegetable plots and chickens go well together. Whilst any outer leaves of cabbages and lettuce that have gone to seed could profitably end up on the compost heap, it must be far better (and certainly more environmentally friendly) to let your chickens feed from such material *before* adding it to the compost pile. And, with regard to those little bubbly lesions that you sometimes find affixed to discarded outer brassica leaves – they are, more often than not, an interim home to the

Just as soon as you've rotovated your vegetable patch for the season, your chickens will be there to inspect!

eggs or grubs of potentially destructive attackers of your next crop of vegetables. If you were to merely discard them on the compost heap, the heat and ideal conditions thus generated might well result in 100 per cent pupation the following spring; if, on the other hand, your chickens were given the opportunity to explore the leaves first and denude them of whatever they find, it's a pretty safe bet that almost nothing harmful will be left to survive.

Should you want to be really clever with your vegetable garden and chickens then look no further than this scenario. In America, there is a system of keeping poultry called 'chicken tractors' – named primarily, I think, after the title of a book written by poultry-keeper Andy Lee. Put simply, chickens are kept in moveable coops and runs and once they've fertilised the ground, the soil is either used to grow vegetables, or the land is kept as improved pasture for other forms of livestock. There can be no doubt that, as well as improving the soil fertility, clean conditions and a more natural diet produce happier birds, more eggs and, should the heart allow, an extra-tasty carcass for the table.

Even if you have neither the space (nor the inclination) for a vegetable patch, your flower garden will – provided that you exercise a little care as to how much free-range access you allow your birds – benefit greatly from their attention. Whilst you might not want your heather-beds turned into a passable impression of the moon's craters as a result of their enthusiastic and energetic endeavours to create a dust-bath, chickens are superb at ridding the garden of grubs and parasites known to be detrimental to ornamental flowers and shrubs. As an aside, you might even get your children interested in digging over the flower beds on your behalf as, whilst they would undoubtedly moan if given the chore for its own sake, it's my experience that they find a great deal of sadistic pleasure in forking over soil and offering whatever worms and grubs they can find to any attendant chickens!

CHICKENS FOR CHILDREN

It seems that a great many chicken-keepers start at a fairly young age (as Clarissa Dickson Wright recently remarked, '...I'm curious about the fondness that the young seem to have for poultry'). The Poultry Club of Great Britain offers several initiatives to encourage active participation by junior members. The Junior Fancier of the Year Award, for example, is an annual competition, of which the final is held at the National Championship Show. To qualify, a junior fancier must have already won

Children and chickens seem to have an affinity with one another – two young prize-winners at a show

an award at one of the shows qualifying for the Championship and, at the final event, not only is the youngster's exhibit judged on its conformation to breed standards, but their poultry knowledge and handling competence are also assessed. The Poultry Club also encourages its young members to participate in a Certificate of Proficiency in Poultry Husbandry – an examination which can be taken by children under the age of sixteen. They claim that the 'syllabus is ideal for schools or community groups'.

Birds particularly suited to being kept as children's pets are considered in Chapter 4, but it is appropriate to include here the possible reasons why some schools are nowadays fighting shy of including chicken-keeping as part of their curriculum. At one time, many schools had rural studies groups, but now, perhaps because of the dreaded health and safety rules, it seems that they are very much in the minority. More need to be encouraged – and they may be after taking inspiration from the thinking behind 'Funky Free Rangers', a project started in 2003 at Staverton Church of England Primary School in Northamptonshire. According to the head teacher, Tim Nelson, 'The main stumbling block for teachers wanting to keep hens always appears to be health and safety. I wish we charged for copies of our risk assessment! There are all sorts of myths and, quite frankly, some absolute rubbish being spoken about out there resulting in hens not being kept in schools because of the supposed dangers. Basically, where we keep our hens, people can avoid them if they have any allergies or phobias. Our children are taught the basic rules of hygiene and we make sure we follow Defra advice on diseases and selling eggs. And, so far, no child has been hospitalised by an angry hen, contracted bird flu or had to be de-loused. And the hens haven't suffered at the hands of the children either, or caught the mumps.'

The original purpose of the project was to give the children some responsibility and the opportunity for them to learn where their food came from. To raise money for the initial outlay on house, equipment and hens, 'shares' were sold to parents and school governors who were promised a percentage of the eggs as a dividend on their shares. The children organised themselves into teams responsible for sales, egg collection, feeding and mucking out, and every decision was reached by means of a majority vote (democracy in action!). As the business progressed, the students – after purchasing a couple of ex-battery hens (now known as the 'Re-Chargeables') to run with their mixed flock of rare breeds and hybrids – learnt all about intensive farming methods and used the necessary numeracy techniques as part of their mathematics class. Pupils at Staverton C of E primary school should, on that basis, know as much as anyone about the economics of chicken-keeping!

ECONOMICS AND COST

All this attention does of course, affect prices and long gone are the days when it was possible for anyone to pick up a trio of extremely well-bred chickens or bantams for fifteen pounds or so. Prices notwithstanding, newcomers to chicken-keeping tend to be divided between those who want rare breeds in the hope of perpetuating a particular type and those who simply require hybrids to lay eggs. On reflection, there is, perhaps, a third element – those people who 'rescue' ex-battery or intensively kept poultry after their commercial life is over. Unless you are a commercial producer, you are never likely to make money from your birds – and why should you expect to do so? It is a hobby after all, and there are very few people fortunate enough to make a profit from a pastime.

Pure-breeds are generally more expensive to buy than hybrid birds, although it is interesting to note that a short while ago, a well-known chain of garden centres was selling hybrids at between twenty-five and thirty pounds each in order to cash in on what they called 'the suburban boom in keeping chickens'. Until recently, a few herb plants and a packet of tomato seeds used to be the main culinary produce you would expect to buy at such places but now at least one has begun selling hens for those who like their eggs as fresh as possible. The company says the move has been driven by the growing numbers looking to live the 'Good Life', and that because chicken coops had been one of their top twenty best-selling items, they had decided to sell the birds to go in them. Sadly, they also claimed that hens will 'happily feed on kitchen scraps, boosting their environmentally friendly credentials' – which is a totally erroneous and misleading comment because chickens of every kind need to be fed a well-balanced and nutritious diet.

Pure-breeds, such as these German Langshans, are generally more expensive to buy than hybrids

Putting the varying costs of purchasing birds to one side for a moment, it would be a poor book that didn't warn of how much your proposed indulgence is going to set you back financially. The first outlay must be in some form of housing and this will undoubtedly be your biggest expense – unless you are fortunate enough to have an existing building that can be adapted. It might be a garden shed, a disused dog kennel or an old stone wash-house, but never be tempted into considering the greenhouse left at the bottom of the garden by previous occupants because a greenhouse is too vulnerable to extremes of temperature and your birds would smoulder in summer and freeze in winter. Generally, though, a specifically designed house and run is your best option. If you are of a DIY nature, you could perhaps take a sneaky look at the design of one for sale at your nearest supplier and build your own, but even so, the materials will not be cheap. The commercially available house and run may well end up being the most economical option as manufacturers have access to bulk buys of timber and workshops permanently set up with pre-formed jigs and templates. Exactly how much a house will cost depends entirely on whether you are happy with a basic, but perfectly satisfactory, design which has stood the test of time, or whether you wish to make a feature of your chickens – in which case the sky's the limit and it's possible nowadays to buy houses made to look like gypsy caravans, lakeside lodges or even a chestnut shingle-tiled beach hut.

If you want to make a chicken run, you will need wire netting – and wire netting costs money. Although you might take a sharp intake of breath when you see its price, it is always better to buy a roll of good quality wire rather than thinking you are saving money by purchasing a roll of wire that has been imported. Almost invariably, the latter is not of the same quality as the former, which should last nearer twenty years than ten. Wire netting will not hold itself up, so remember to include the cost of posts and incidentals such as straining wire and even netting staples.

Despite various factors affecting world cereal prices, food, whilst not exactly 'chicken feed', is relatively inexpensive – and, with say half a dozen hens of a known egg-laying type, the sale of their eggs to friends and neighbours may even cover that particular cost (as a hobbyist, you do not need to register as an egg producer to sell eggs to such people). You will, of course, need to buy feeders and drinkers: as with the wire netting, it could prove false economy to buy any that are not well constructed by a known, reputable manufacturer. Eventually, as you become even more enthusiastic and knowledgeable, you might feel the need to shell out (forgive the pun) on a little incubator and brooder, or even a short row of penning cages in which you can train your exhibition birds in preparation for shows. Nevertheless, not everything has to be bought at the outset and it is

Chicken houses can be decorative as well as practical

usually a case of equipping oneself with the essentials before deciding on the luxuries.

Even with a cautious start, sadly, pennies do have a habit of becoming pounds. However, one should not turn a blind eye to the need for incidentals such as floor litter, anti-parasite medications, leg-rings (useful for identifying individual birds or siblings from the same hatch or strain), subscriptions to your local poultry club and chicken-orientated magazines, or even something as seemingly inconsequential as a notebook in which you can record financial out-goings as well as day-to-day details appertaining to your flock.

▲ 'Incidentals' do add up – membership of your local club, subscriptions to a magazine and entry fees to a show all cost money; a notebook or record of expenditure might prove interesting at the end of the year! *(Author's photo)*

CHICKEN SCHOOL

For those who might have missed out on the possibility of chicken-keeping at primary or secondary school, it is, thankfully, never too late to learn the reasons why it's a good idea to keep chickens – and, perhaps more importantly, what one has to do to ensure a modicum of success. Before embarking upon ownership, it might be wise to consider attending a basic poultry husbandry course at your local agricultural college or independently run farm school. There are plenty of opportunities on offer, many of which operate on a half-day, day or weekend basis. Details of up-coming courses are usually advertised in chicken-orientated or smallholding magazines and you can usually find relevant website details without too much trouble. It is also worth checking to see whether your local library has any information.

Typically, the subjects covered include (depending upon whether the course is intended for absolute beginners or the more experience chicken-keeper), the different types of breeds, where you can buy them,

housing and runs, feeding, disease, worming, dealing with mites and lice, protecting your chickens from the fox, introducing new chickens, as well as handling chickens and clipping their wings. More advanced courses might briefly cover all or some of the above and then, in addition, go on to discuss welfare issues, legal aspects, predator control, the pros and cons of incubators and broody hens, organic poultry-keeping, and even the slaughtering and plucking of chickens and preparation of the carcass for the table.

Many chicken businesses organise Open Days where you can get some idea of the types of chicken you might like to keep; it should also be possible to see several examples of different housing systems and the equipment necessary. In addition, there are places like The Domestic Fowl Trust to visit. Situated near Evesham, Worcestershire, the Trust nowadays also incorporates Honeybourne Rare Breeds, and the entry fee entitles visitors to look round all the livestock on display, not just the poultry. As well as the paddocks full of chickens and bantams (some of which are kept in mock-Tudor-style houses!), there is a Defra-funded museum dedicated to displaying some fascinating poultry artefacts, such as old incubators, egg-graders and all manner of information from which the newcomer to chicken-keeping will surely learn.

'Proper' school and 'proper' lessons!

With any of the above, there is, as with most things, a cost involved – although I notice that one particular establishment does say that, if you live within its catchment area, the Duchy of Cornwall will offset your costs up to 50 per cent!

If you want to get technical, there are sometimes full training grants available – but it must be admitted that the majority of the poultry training courses for which these are applicable are intended for farm workers and managers rather than new entrants and hobbyists.

At the Domestic Fowl Trust, Evesham, Worcestershire, some of the chickens and bantams are kept in rather fine Tudor-style housing!

However, I include the information below just in case a hobby gets out of hand and you, or a member of your family, decide to make a career change! As Andrew Warriner, a free-range egg producer from Pickering, North Yorkshire, said after completing his course, 'The courses are fantastic. They've given me much more insight and knowledge about hens and this is helping to increase production and tackle any issues with the flock.'

Generally, the courses are designed to help would-be poultry producers gain and then maintain accreditation, such as with the Lion Code (eggs), RSPCA Freedom Foods (eggs/broilers/ducks), Assured Chicken Production, and Quality British Turkey Assurance. The courses cover veterinary health, biosecurity, welfare, vaccination and medication, laboratory procedures, housing and the environment, egg production, food and supply chain, insect and pest control, vermin control, fire awareness and health and safety. I suspect, whilst of undoubted interest, this may be a stage too far for the average back garden chicken-keeper who wants nothing more than to learn where and how to look after up to half a dozen birds on a daily basis! Mind you, as I mentioned in the previous paragraph, who knows just how bad the chicken-keeping disease may take hold – that it is a disease there can be no doubt, as Rosslyn Mannering, writing in *Fowls and How To Keep Them* (published in 1924) observed in her book's preface:

> ‘IN WRITING THIS LITTLE BOOK on poultry I have had in mind chiefly the needs of the individual who wishes to be independent of the shops for eggs and, perhaps, table-poultry supplies. But since 'hen-fever', when once contracted, is a complaint that usually increases in virulence, I have also kept before me the possibility of an expansion in the original venture of half-a-dozen layers, and indicated the lines along which the enthusiast may progress successfully and within the scope of his accommodation.’

3

WHERE TO KEEP CHICKENS

The correct answer to the question raised in the chapter title is, 'virtually anywhere'! The great thing about chickens is that they can, more or less, fit in to whatever space and location is available and there is almost certainly a breed or type that will prove suitable for the situation in which you find yourself. However, chickens cannot, and must not, be kept in cramped and over-crowded conditions and it is the legal responsibility of any chicken-keeper to ensure that their birds are correctly maintained and humanely housed – especially in the confines of a built up area where there are neighbours to consider. Vermin in the shape of a few rats and mice might not pose too great a problem at a sequestered rural cottage (and can be controlled with safe baiting-points), but to allow the same in an urban back garden is a good way of

A cock bird will lose no opportunity to declare his presence and will use any vantage point to do so!

receiving complaints and ensuring a visit from the Environmental Health officials. A cock bird's constant loud crow is not likely to endear either him or you to the neighbours and if things got really out of hand, notice to remove the bird could be served on you by the local council.

Neighbours notwithstanding, and assuming that the majority of would-be chicken-owners reading this live in a suburban environment rather than out in the wilds, most back gardens are sheltered enough and suitable. The positioning of the hen house and run will, in all probability, be limited by the garden layout, but if you are fortunate enough to have plenty of space, try and choose a place where birds have the benefit of any early morning sun. Constant winds are to be avoided so pick as sheltered a spot as circumstances will allow, but, conversely, do not worry that, just because a winter's day might seem cold (rather than windy) to you, your birds will feel the same because provided they can find a little shelter from the worst of the weather, they will be perfectly fine. Do, however, bear in mind that, although it might be supposed that surrounding buildings and gardens will offer a degree of protection, depending on their layout, they might actually create something of a wind tunnel. If the chicken's home and run can be kept along the shelter of a boundary hedge or maybe under the shelter of a specimen tree, not only will this help protect against frost in the winter, in the summer months they will also have all-important shade from the sun (chickens can suffer badly from heat exhaustion).

Of course, if your garden is totally secure, it might only be necessary to give chickens a night-time roosting place and otherwise allow them free range, in which case (unless you have absolutely no interest in your garden), it could just be a question of fencing your chickens out rather than in! Both large fowl and bantams (especially bantams) like to explore and will take great delight in discovering the most interesting tit-bits inevitably to be found in either a flower border or the vegetable patch, so a few neatly erected pieces of aesthetically pleasing lattice or framework around such places could be required if you are not to be continually frustrated by their inquisitive exploits.

What's legal – and what's not?

If you look at any Internet poultry forum, one of the most frequently asked questions is whether there are any laws to prevent anyone keeping chickens in the back garden. The answers are many and varied, but the 1950 Allotments Act loosely states that if you have access to land as your own then you are permitted to grow vegetables and keep rabbits and chickens for food purposes (eggs are undoubtedly food!). Even so, not all councils will let you keep chickens on allotments rented from them, and tenants in rented accommodation (either council or private) must at least check with their landlord.

It is important to remember that regional by-laws are not necessarily the same as a national government Act and there may well be some local clauses that prevent chicken-keeping. Even if there are not, you should be aware of other legislation which may be applicable, for example, keeping chickens in the garden is legal as far as planning laws are concerned, but, while there are provisions in the Planning Act to construct or have chicken sheds in the garden, there may well be height, size and positioning restrictions.

A moveable combined house and run has many advantages over the traditional poultry pen

Freedom-fighters

Before considering keeping chickens you should be aware that legally, one has a 'duty of care' towards supplying them with their needs, or the five 'freedoms'.

- **Freedom from hunger and thirst** – by ready access to fresh water and a diet to maintain full health and vigour.
- **Freedom from discomfort** – by providing an appropriate environment including shelter and a comfortable resting area.
- **Freedom from pain, injury or disease** – by prevention or rapid diagnosis and treatment.
- **Freedom to express normal behaviour** – by providing sufficient space, proper facilities and company of their own kind.
- **Freedom from fear and distress** – by ensuring conditions and treatment which avoid mental suffering.

HOUSING

In reality, the most suitable home for chickens will, in the smallest of garden areas, probably be a moveable combined house and run which, it has to be said, has many advantages over the traditional poultry pen as it is likely to be warmer, afford better protection from the elements, allow the inhabitants semi-freedom (they can always be let out for a potter round the garden when someone is on hand to see to them) and offers security from predators. Perhaps most importantly, however, the whole thing can, from time to time, be moved onto fresh ground – thus ensuring that there is less chance of a build-up of parasites or that the lawn becomes too 'scuffed' as a result of their enthusiastic scratching. Normally made of traditional timber, the size and design of these units are many and varied and to see evidence of this, one only has to look at adverts in magazines such as *Your Chickens, Smallholder* and *Fancy Fowl*

Covered runs have several advantages over open-topped ones. At a show, it's always possible to get advice as to which will best suit your particular situation

Buildings for tomorrow

While planning the housing, it is as well to consider the logistics of obtaining a small extra house and integral run which is normally kept empty. It can have a multitude of occasional uses, including being called into action as a rearing place for a broody and her chicks, to fatten a couple of surplus cockerels, to separate birds from the main flock (for a variety of reasons) and, last but certainly not least, as an 'isolation ward' for stock that looks a little off-colour (it makes good sense to isolate any sick-looking bird until a definite diagnosis can be made). Without being unduly hard-hearted, it is my personal experience that a truly sick chicken rarely recovers, but one that may have a minor injury or has been bullied by the remainder of the flock will, if given isolation and a little 'TLC', almost certainly make a full return to health.

in order to check out the huge range of styles on offer. Many are given glorious-sounding names (who would not want their chickens living in the 'Grand Swiss Chalet', the 'Blenheim' or the 'Regency'?). Some have open wire run attachments whilst others have the roof and possibly even the back of the outdoor run weather-boarded. This will obviously protect the occupants from unpleasant winds and driving rain, but there is another advantage should one be intending to keep and show white or very light-coloured breeds, which, in an open run, may well become discoloured or 'brassy' (*see* Glossary) as a result of constantly being exposed to the elements.

And then, of course, there are the proprietary chicken units which are not made from wood at all, but are constructed of heavy-duty (often recycled) plastic. They have revolutionised back garden chicken-keeping and have, it must be admitted, many advantages over housing made from wood, not the least of which is their hygiene benefits. Unlike timber constructions, these units are easy to clean and can be pressure-washed on a regular basis. Their lightness also means that they can be easily moved around the garden or, if necessary, carried to a neighbour's house so they can look after your stock when you're away from home. Some traditionalists do, however, remain concerned about the possibility of condensation building up on the interior of a plastic shed and also whether such units are truly fox-proof. The manufacturers are, quite naturally, quick to refute both these criticisms.

The real traditionalists might be happiest to see your chickens housed in a wooden shed that stands centre-stage in an open-topped run. They are usually similar to a garden shed in design (in fact, the majority of garden sheds make perfectly good chicken houses) and have the advantage of being tall enough for the owner to walk easily in and out in order to collect eggs or carry out the daily chores. A tall shed is also generally advantageous as far as the health of chickens is concerned as there is more scope for adequate ventilation and air movement; their height does, it must be admitted, mean that there is more scope for heat loss on a particularly cold night.

A modern hen house made from recycled materials – easy to take apart and clean *(Photo courtesy of Green Frog Designs, Somerset)*

It might sometimes be possible to come across a small, old but perfectly serviceable poultry ark at a farm sale or similar. My first ever pen of bantams were housed in an ark and run bought in the late 1960s for the princely sum of ten shillings (fifty pence) from a retired poultry enthusiast who had no further use for it. Transport was a problem so I recruited the aid of three of my school friends and, between us, we carried it the couple of miles home along country lanes, over stiles and across farmers' fields! It had no solid floor in the house part and my father constructed one out of second-hand floor-boards; I made a pair of nest boxes from an egg carrying crate and, together, we shaped and fitted a perch.

Building a chicken house

Although it has been suggested previously that it might be cheaper to buy a ready-made hut and run than to consider making one, there are those, I'm sure, who might still be tempted. For them, I include a few basic principles, first and foremost of which is to consider not only the needs of the occupants, but also yourself. Build it with enough headroom for you to be able to clean it out easily and make sure that you include an access door that doesn't require the skill of a limbo-dancer to get to the farthest corners. Ideally you should be able to open the nest boxes from outside rather than having to walk inside the house every time you want to collect eggs.

Craftsman at work! Ornate steps being prepared for fitting to the 'Gypsy Willow' chicken house model at the Flyte so Fancy workshop *(Photo courtesy of Flyte So Fancy Ltd, Dorset)*

The general construction should be of good quality timber, both for the framing and the cladding. Wood is naturally warm and, if treated properly, will last for many years. Make a point of coating each part with proprietary waterproofing during building as this will ensure that every joint and recess is well protected. After completion give everything a further coat or two according to the manufacturer's instructions and leave to dry thoroughly before housing your stock. In an ideal world, the house should be given a good scrub down with disinfectant, inside and out, once or twice a year and a further coat of preservative applied.

One of the most important requirements in all poultry houses is adequate light and ventilation. Lice and mites survive mainly in places where there are dark corners in which to set up home. A coat of animal-friendly paint on the interior walls will help reflect all available daylight, and it is surprising how much extra light can be gained by the simple expedient of keeping any glass panels periodically washed! Good ventilation is vital in preventing respiratory diseases. This can most obviously be provided by natural means such as windows that will open easily. Other methods include a protected open ridge running along the apex of the larger poultry houses, or even something as simple as a row of holes drilled at the top of the sides just under the eaves or roof overhang. There is, understandably, a great deal of difference between the need for ventilation in the summer and winter months and, in practice, ventilation is only really necessary in winter in order to get rid of stale air when even the hardiest of birds will favour their house as opposed to being out in the most extreme elements.

Make any attached run as long as is practicable – bearing in mind that there will, at some time, inevitably be a need to capture birds in one of its furthest corners. An access door in the top means that it is less likely that the occupants can escape while being fed and watered, but including such a refinement means that you will have to include some sort of pop-hole or small gate in the end or one of the sides if it is intended that your birds will be given access to the garden from time to time.

HOUSE FITTINGS

Whether you are designing your own hen house or purchasing a ready-made unit, the basic needs as far as fixtures and fittings in the house remain the same, though some may need adapting in order to take into account the peculiarities of your chosen breed – while you might be able to 'fit a pint into a quart pot', you would most certainly not be able to fit a quart into a pint!

Pop-holes need to be large enough to accommodate whatever type of chickens you are keeping

Perches

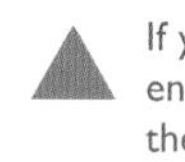

If you are intending keeping birds in large open enclosures, perches could also be included in the run: your chickens will undoubtedly enjoy using these during the day time

In order to keep them away from floor draughts (and to indulge their natural roosting instincts), your chickens will benefit from a perch or perches fitted inside their house. The construction and height will vary depending upon the type of housing and also the breed of chicken. Obviously, heavy breeds cannot scrabble up to a perch which is set high (and, in any case, no perch should be so high that there is the remotest possibility that the birds might hurt themselves either going up to roost in the evening or coming down to the floor again in the morning). Average measurements might, therefore, be a height of about 60–90cm (2–3ft) from the floor – but it does depend on circumstances and, the smaller housing units are often not much higher than this in total so perches will have to be fitted accordingly.

Each perch should be around 5cm (2in) wide, and about 3cm (1in) thick for bantams or small breeds and of slightly thicker dimensions for larger, heavier birds. Any timber used should be planed relatively smooth and, in the opinion of most chicken-keepers, have the top edges rounded slightly in order that your chickens can take a secure grip. If more than one perch is required, make sure that they are all set at the same height, and don't have perches running parallel to the building any closer to the wall than around 30cm (1ft).

Nest boxes

A set of modern-day nest boxes made by Osprey Plastics (BEC UK) from recycled plastic *(Photo courtesy of Osprey Plastics (BEC UK), Shropshire)*

Traditionally, nest boxes are made of wood, but on several occasions I have seen home-made boxes adapted from plastic 25-litre (5-gallon) square drums obtained from a farm, which I thought was an excellent idea because they can be easily washed and disinfected. In fact, there is an all-plastic poultry nesting box available which has the advantage of taking just seconds to clean and will most certainly help in keeping nests free from parasites, pests and bacteria. The manufacturers claim that their particular product 'is extremely easy and quick to clean out, has no corners to harbour mites and can be jet washed in seconds, ending the hours of drudgery cleaning and disinfecting wooden or metal boxes to ensure they have no red mite or bacteria in them. Unlike wood and metal [it] does not corrode or rot, it is cheaper and at the end of its life it can be recycled.' They can, apparently, be fitted into the hen house double stacked or placed in line and also have an optional egg tray for ease of collection. Whatever the choice of nest box, you should ensure that there is one for every three or four hens you keep.

It is important that nest boxes are regularly cleared of any faeces, and fresh litter added so that eggs are kept clean; this will prevent the staining of eggs and possible bacterial infection entering the porous shells. The nesting material can be the same as that used on the floor, but I would not recommend sawdust for either because of the high dust content – nor would I suggest hay, as it tends to harbour parasites. Even as a child, I somehow knew that, pretty though it looked and as aromatic as it smelt, a nest box full of hay was not the cleanest or most hygienic of litter as, unless it is perfectly cut, dried and meticulously stored, it is potentially subject to fungal diseases. If you do choose to use straw, use only the wheat type.

Floor litter

As is the case with the nest boxes, suitable floor litter for the house itself is very important. Its main purpose is to aid cleanliness and hygiene, and traditionally would have been straw or wood shavings; however, the upsurge of interest in chicken-keeping has seen the development of many types of new poultry litter. Some are just refinements of old ideas and take advantage of the quite understandable wish to used recycled materials, whilst others are totally revolutionary, and it is therefore well worth taking time to do some Internet research on what is currently available. Flax in its various forms appears frequently on poultry suppliers' websites and is extremely absorbent; it is slightly more expensive than some other materials, but the extra cost is outweighed by the fact that less is used. There is also much to be said in favour of chopped hemp as both floor and nest box litter. It is a natural, dust-free, sustainable, eco-friendly product which will absorb up to twelve times more liquid than straw and four times more than shavings. The fact that it is quick to break down makes it of great value in a back garden where soiled litter is added to the compost heap before eventually being recycled to the vegetable patch.

Fanciers of feather-legged breeds such as these d'Uccle Belgium bantams often advise the use of sand, either on its own or mixed with a little gravel, as a floor covering

Chicken compost

It might just be worth a quick word here regarding chicken manure: it contains, as is well known, far greater levels of nitrogen than any other type of livestock manure and must therefore, never be added straight to the vegetable garden. Too much nitrogen applied directly can burn tender seedlings and cause more established plants to create lush sappy growth which is then susceptible to attack by pests and disease. Good though nitrogen is for the soil, too much may create a situation where the only plants able to cope are nettles – not at all desirable in the average back garden!

So, always put chicken manure onto the compost heap first, where, because of its brilliant 'activating' qualities, it helps to break down all the other vegetable matter which normally finds its way onto the heap. If it is put in there together with soiled wood shavings from the floor of the house, the nitrogen it provides will compensate for the negative qualities of the wood shavings which are infamous for robbing the soil of this essential nutrient – in effect, one will complement the other.

CHICKEN RUNS

Some of the heavier breeds of chicken have, over the years, developed in such a way that their wings are hardly capable of doing much more than flapping and lifting the bird a few inches off the ground when danger threatens. Others, usually the lighter Mediterranean varieties, can fly quite well and are certainly capable of getting into the branches of a nearby tree if pursued by a domestic dog or foraging fox. Logically then, it should be easier and cheaper to build an open enclosure suitable for the heavier breeds such as Sussex or Wyandottes rather than one for Minorcas or Anconas – and so it would be if all one had to consider was their escaping. In reality, though, an outside run to which birds are going to be allowed permanent access also needs to be a safe and secure structure to prevent predators from getting in. Obviously the size of enclosure is dictated by the number of chickens you wish to keep, but it is always worth making it larger than you initially think in order that it doesn't become a muddy unpleasant mess during inclement weather. Just because you have more space, don't be tempted into filling it with more birds at a later date – I make no apologies for reiterating the fact that chickens and bantams do not do well when kept in overcrowded conditions.

Preparing to clip a chicken's wing in order to prevent it flying over the wire of its run

Yes, believe it or not, this is a chicken house! Why not make a feature of yours rather than hide it away in the corner? *(Courtesy of Flyte so Fancy Ltd, Dorset)*

To protect against foxes (and remember that a lot of foxes nowadays live in towns and suburbs) or even wandering dogs, the fencing surrounding permanent poultry buildings should be around 2m (6½ft) high, with at least 30cm (1ft) dug into the ground (or turned out and pegged) to prevent anything from scratching under. Personally, I prefer to make the turn out and first 0.6m (2ft) of the fencing from small-mesh wire netting which adds strength and also helps to protect the run from rats, stoats and even hedgehogs (who are quite notorious for

stealing eggs!). The remainder of the height can be made up of larger sized mesh. Strain the tops and middle sections with a length of galvanised wire stapled to the posts, but do not hammer the staples right 'home', or they will abrade the wire and cause it to break.

Use strong posts: angle-iron ones with holes drilled at appropriate heights to allow a length of straining wire to be threaded through work well and will last a long time, but most people tend to use wooden stakes – which can be obtained either from your local agricultural suppliers or a nearby wood-yard. They will, in all probability, have been pressure-treated with a preservative which will help with longevity.

FUNDAMENTAL OR FEATURE?

It will, I hope, have been made obvious that there is no need to hide your chickens away down the bottom of the garden in a shabby house made from disused packing cases – although, having said that, I once built myself a very serviceable bantam hut from a packing case used to deliver parts of David Brown tractors to their Meltham factory. Instead, make a feature of wherever it is you choose to keep your chickens in. It might not even be too ridiculous to suggest that there should be climbing roses round the door (why not?) or that your chicken accommodation takes centre stage in a quite ornate form. As will be seen later (Chapter 8), the Victorians and Edwardians gave due prominence to their poultry – so why shouldn't you?

Yes, chickens will undoubtedly do well in a healthy utilitarian environment but, if finances allow, there is absolutely no reason why they should not be given their own chicken château. Who knows, should the day ever come when you want to move on, such facilities might even add to the value of your home. We've all watched reality TV shows in which someone wants to relocate or move to the country and have a vegetable patch and keep chickens – if your premises are such that all is already in place, it might just be the thing that clinches the deal.

4 WHAT CHICKENS?

Large, small, colourful or dowdy, light or heavy in conformation, there is a type of chicken to suit everyone

Having a clear idea exactly why you want to keep chickens and where you can do so is likely to make the choice of chicken breed considerably easier. Some writers do, however, tend to confuse matters by suggesting that the likes of 'Legbars' and 'Black Rock' are a breed – they are not; they are a hybrid type and will not breed true. They are, though, excellent egg layers and there can be no doubt that there is a most wonderful simple pleasure to be had in collecting a freshly laid egg from the nest box, no matter what chicken it came from. The colour of the egg should be immaterial, but when it comes to placing a soft-boiled one on the breakfast table, the fact that it is brown certainly adds to the attraction. Is the colour of its eggs really sufficient reason to choose a particular breed or chicken type? – a lot of people think so and it is, as a result, one of their main criteria when deciding what breed or type of chicken might best suit them.

Large, small, colourful, relatively dowdy, light or heavy in conformation, there is a type of chicken to suit everyone, whether it is their intention to keep them for their eggs (brown or otherwise), as an enhancement to the garden, or in order to try their hand at showing and exhibition. In the latter case, the birds you choose will obviously need to be a pure-breed recognised by the Poultry Club – but more of that in Chapter 8.

DEFINING 'RARE BREED' STATUS

When a bird is classified as a 'rare' breed, it does not necessarily follow that actual numbers are scarce (at least not in the UK – other countries categorise their breeds differently). In Britain, it might be that, rather than being rare in overall numbers, a breed is rare in its geographical concentration – as the Rare Breeds Survival Trust (RBST) points out, 'some breeds may be numerous but if the majority [of birds] are found in a small geographical area the breed will be highly vulnerable to disease epidemics.' Their categories of 'critical', 'endangered', 'vulnerable' and 'at risk' are therefore likely to relate to this factor.

With the publication of its 2011 *Watchlist*, the RBST has announced a new separate listing for poultry which recognises the need to take a

An entrant in the 'rare breeds' class being judged at the National Show

different approach to assessing rarity in poultry breeds as compared to other farm livestock. The new listing is a result of the findings of a working group set up in 2010, and 'UK Poultry Breeds At Risk' will stand alongside the standard *Watchlist* for cattle, sheep, goats, pigs and horses. Lest you query their credentials, the RBST Poultry Working Group consists of representatives of the Poultry Club of Great Britain, the Rare Poultry Society, the Turkey Club, and individual poultry specialists. Looking at poultry in the context of the *Watchlist,* the Group concluded that it is not possible to apply the same criteria as for other species.

As RBST Conservation Officer Claire Barber says: 'While there are many breed societies for poultry, there is no single central registration system and it would be very difficult to establish one. This has resulted in the total numbers of these breeds always being very difficult to estimate. It had become apparent that it was pointless trying to shoehorn poultry into the standard *Watchlist* process. However, this does not mean that poultry is not considered as important as other *Watchlist* breeds. On the contrary, it is because of the importance of poultry in livestock systems that we need to have a more appropriate poultry listing.' With the listing revisions now made, the Poultry Working Group will continue to review poultry issues for the RBST. It has already added a number of new breeds to the existing list and it describes the UK Poultry Breeds At Risk document as being 'a work in progress'. The 2011 UK Poultry Breeds At Risk listing can be found on *www.rbst.org.uk* together with the guidelines for acceptance of breeds onto the list.

MIXING BREEDS TOGETHER

There are so many beautiful breeds of chickens around that it is very tempting to make up a pen containing a wonderful collection of half a dozen or more colourful birds. Indeed, I have seen this advocated in several chicken-keeping magazines and also on Internet blogs and websites. In theory, the idea is a lovely one – and for a multitude of reasons, the most obvious of which is to be able to enjoy the

Lovely though a mixed pen of birds may be to look at, different breeds do not necessarily go well together

characteristics of several breeds all at once. Plumage variations add to the interest and there is, of course, the fact that individual birds of different colours and breeds are so much more readily identified and can even be given appropriate names: 'Winnie' (or 'Winston') the white Wyandotte, 'Lettie' the lavender Leghorn or whatever.

Great care must, however, be taken in ensuring that breeds are compatible and able to live alongside one another. Perhaps the best example of this is the Brahma (Orpingtons and Faverolles are two others). Amongst its own kind, it is the kindest and most gentle of breeds; in truth, it is sometimes too kind and gentle for its own good and, if you were to include it in amongst a collection of mixed breed, you might well find that, despite its enormous size, it tends to be bullied and, as a result, will very soon become submissive and extremely unhappy.

Whilst it can reasonably be expected that tiny bantams are quite likely to be set upon and bullied by any of their large fowl cousins, the opposite is actually more probable. A particular breed might well, amongst their own kind, follow you happily round the garden and make the perfect pet as far as human compatibility is concerned, but set loose alongside other chickens, their tenacious and pugnacious attitude may well come to the fore.

Should you wish to keep both large fowl and bantams of the same breed together, there is, apart from the natural establishment of a 'pecking order', usually very little to worry about because the general characteristics of the breed normally show themselves; therefore, if the breed is known for its suitability as a practical, pretty or pet bird, whether it is big or small matters little.

PRACTICAL, PRETTY OR PETS?

It has never been my, or the publisher's intention to make this book a comprehensive dissertation on chickens – merely to offer the reader some sound general advice which should go a long way towards ensuring that they achieve a modicum of success with their chosen hobby, and, more importantly, that they will obtain the maximum amount of pleasure from it. Bearing this in mind, it would, therefore, be out of place to delve too deeply into the many and varied pure breeds of birds that are available. I recommend instead that, should anyone be searching for such a list, they arm themselves with a copy of one of the several books currently available that have 'choosing chickens' as a part or whole of their title.

I nevertheless thought it might prove useful if I made reference to just a few of the more commonly known and readily available breeds. In the main, my choices cover the less complicated or potentially less troublesome breeds. I have split this section into 'practical', meaning those suitable for egg-laying or possibly even the table; 'pretty', suggesting that although they might not be good egg layers, they are an attractive addition to the garden (and might also make good exhibition birds); and finally, 'pet' breeds which are calm and friendly (especially with children in mind). Of course there is absolutely no reason why some breeds mentioned cannot be practical, pretty *and* make perfect pets!

PRACTICAL

One of the greatest advantages that pure-breed (sometimes known as 'pure-bred' or 'standard') chickens have over cross-breeds is that it is possible to predict their adult size, conformation and likely behavioural characteristics, all of which are important factors in deciding whether or not they will be suitable and practical in your particular situation. Any of the following breeds are quite hardy and undemanding in their needs so should prove a good choice for the beginner.

Leghorn

The Leghorn has its origins in Northern Italy and is seen in several colour variations, perhaps the two most common of which are the black and the white. In the UK, brown, buff, blue, 'exchequer' (a sort of mottling of black and white), partridge and even multiple pencilled partridge (which is a bit of a mouthful and might perhaps be more easily called 'triple-laced') can also be found. No matter what colour, they are all extremely handsome birds, and the cock bird normally has a large fairly erect single comb (although it is sometimes possible to see rose comb varieties) while the comb of the hen should flop over to one side – but not so much that it obscures one eye. It should perhaps be mentioned that it is just possible these combs might succumb to frostbite in very severe cold weather (*see* Chapter 6). Exhibition strains of the breed tend to be a little larger than those bred simply for their egg-laying value – and they are one of the best laying breeds around – but their light conformation means that they can be slightly more flighty than some other breeds and you may need to keep them in an enclosed pen if you don't want them forever exploring the innermost parts of your garden! They rarely, if ever, go broody and will lay eggs regularly on almost a daily basis, but even a prolific laying breed such as the Leghorn will, at the end of the summer or in early autumn, need to take some time off in order to go through the moult (*see* Chapter 6). They are relatively easy to keep, and their eggs are white and of a good size, so the breed is quite near the top of the list as regards the practicalities of chicken-keeping.

White Leghorn male

Rhode Island Red

Although it originated from America, the Rhode Island Red is nowadays extremely well known in Britain. It is the classic chicken, as often portrayed in children's books, and its feathers should be a lovely deep colour rather than the wishy-washy brown normally associated with many hybrid types. Some might, however, still say that it is relatively plain to look at, but beauty is in the eye of the beholder and this good, solid, straightforward breed takes some beating when it come to ease of care.

Despite its origins, it wasn't long before the Rhode Island Red was exported to many parts of Europe where it was used to cross breed with other chicken types – as a result of which their genes are to be found in many of today's commercial hybrids. They are also interesting from the point of view that, in the mid-twentieth century, they were often mated with the Light Sussex because, depending on which sex of which breed was used, such a pairing produces chicks that can, from their colour, be immediately sexed within hours, rather than weeks, of hatching. This saves producers of free-range eggs the time and expense of rearing a batch of chicks to a certain age before they are able to cull or sell unwanted cockerels. The practice is known as 'sex-linking' and there are several other breed combinations with which this can be done. The eggs of the Rhode Island Red are light brown in colour, and there is also a white variation of the breed, but this is far less commonly seen – despite the fact that both lay a comparable number of eggs (although there are some who will tell you that the white variety lays fewer).

Rhode Island Red bantam male

Sussex

Along with the Rhode Island Red, the Sussex, and particularly the Light Sussex, has been a favourite among both farmers and back garden poultry-keepers in the UK for many years, particularly during World War II when it was kept extensively as part of the 'war effort', providing both eggs and meat to households which would otherwise be lacking in

Light Sussex female

such commodities. Originally the Light Sussex was primarily a table bird, but with subsequent breeding the meat production aspects of the bird's make-up have been lost in favour of a lighter, egg-laying breed and there are both hybrid-sized and much larger traditional types to be seen. The plumage variations are, in some types, difficult to breed to show standards, but none of the colour types is any less easy to keep. The Light Sussex markings (predominately white with black in the hackle, end of the tail feathers and edges of the secondary wing feathers) are, in other breeds of chicken, sometimes referred to as 'Columbian'. The breed has a single comb and the hens are excellent layers of cream to light brown eggs. The birds are full of character and if you are looking for the traditional 'farmyard' bird, your search is over!

Marans

Well known and loved in Britain, the Marans is still hugely popular in its home country of France, and members of the breed club there go to great length to ensure that examples of the breed are bred not only for conformation, but also that future generations will continue to lay eggs that have a 'deeply dark-brown coloured shell'. In fact, for many French breeders, the egg colour is probably the most important factor and they will not breed from a pullet until she has laid several eggs and the colour is proved to have remained consistent. A hen that lays a single dark chocolate-coloured egg and then reverts to those of a paler pigment most likely does so due to an abnormality of glandular secretions as the egg passes through the oviduct and would not be considered suitable for inclusion in the French breeding pen.

Sometimes known as 'Cuckoo Marans' because of the colouring most commonly seen, the breed can (depending on what part of the world you find yourself in) be seen with colour variations known variously as black, copper black, white, golden, silver-cuckoo and Columbian (a little like the colour markings of a Light Sussex). Ideally, the comb should be of the single kind and possess five or six points. The breed is well known for its friendly character, making it another ideal bird for the beginner – in fact, when we had Marans running around the garden, on sunny days when the back door was open it was a frequent occurrence to find them inside the kitchen ready to carry out a conversation with any human, cat or dog who was prepared to listen!

Cuckoo Marans female

Welsummer

Other breeds of easily maintained hens that lay deep brown eggs include the Welsummer, which originates from Holland. Although it might not lay quite as many in a season, their egg colour is comparable to that of the Marans (the only real difference being that the shell of the latter is glossy whereas that of a Welsummer is matt). Generally only ever found in one plumage variety, the cocks being a pretty mix of red, brown and black, the hens brown speckled with gold-tinted hackle feathers, there is also a silver duckwing colour. The breed (large fowl or bantam) is, appropriately enough for this section, very practical because it is friendly, attractive to look at, adaptable to most conditions and extremely hardy. Being a little slighter in conformation than its large fowl counterpart, the bantam version is known to be livelier and more active, but it is no less suitable in terms of domesticity and character. Welsummer bantams, compared to their large fowl equivalent, which was developed at the end of the nineteenth century, have only been in existence for sixty or seventy years and were created by using small examples of Welsummer hens crossed variously with Rhode Island Red and Wyandotte bantams.

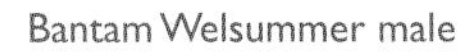

Bantam Welsummer male

It has been said that the breed's only downside is that the hens make poor mothers; why this should be, given that they are a reasonably heavy and placid breed (and especially bearing in mind the fact that the

bantam version has bloodlines from two breeds well known for their brooding instincts), I must admit to having no idea. It is a point perhaps best raised with an experienced Welsummer owner – any one of whom will, I am sure, give you more detailed advice.

Wyandotte

I could wax lyrical forever about this most wonderful breed of chicken. Having said at the outset that it was not my intention to include the breeds that, for one reason or another, might be too difficult for the inexperienced chicken-keeper to contemplate, the various Wyandotte types illustrate perfectly the varying degrees of knowledge one might require in order to build up and maintain quality breeding and potential exhibition stock. The good old-fashioned white must be somewhere at the top in ease of breeding, making it the perfect beginner's bird. The black variety is perhaps somewhere in the middle with its tendency for recessive genes to throw chicks with wrong-coloured legs, but the partridge coloration is a veritable minefield, needing separate pens of cock and hen breeders in order to accommodate their complicated plumage patterns. Putting all that to one side, the breed is the 'number one' at the top of many chicken-keepers' lists of favourites, and, with its attractive rounded shape, good egg-laying ability (some colours will lay better than others) and perfect personality, it is easy to see why. Its attractive appearance is further enhanced by a bright red rose comb which has (or should have) a 'leader' at its rear, the line of which should follow the head and neck line rather than stand proud. There are at least fourteen possible colour types recognised as standard in the UK, but worldwide there are many more. The egg colour is usually referred to as cream, but, depending on the particular strain and type, could in fact vary between tinted and almost white.

If space and time permitted, I would have no hesitation whatsoever in including at least half a dozen varieties of this most wonderful American export in my own garden – and, not content with that, I'd have examples of both large fowl and bantams!

▲ Blue-laced Wyandotte female

The Best of the Rest

While the above are considered by most to be the perfect breed choices for the new or relatively inexperienced chicken fancier, there are plenty of other breeds that might suit almost as well. Here are brief details of another five possibilities.

Australorp

The breed is popular in the UK (2011 was the ninetieth anniversary of the Australorp Club of Great Britain) and gets its name from an abbreviation of 'Australian Black Orpington'. Australorps were created by commercial producers from the Orpington breed which had been taken to Australia from England. The expression 'coals to Newcastle' springs to mind! Originally it was bred as a dual-purpose bird and as well as providing a good meat carcass, the hens will lay upwards of two hundred tinted eggs a year – so it has good reason to be at the top of the 'best of the rest' list. It is almost always black in colour, but there is also a white and a blue-laced variety. It is practical and hardy; being heavy in shape and conformation, it is not at all flighty and quickly becomes used to people and the back garden routine.

Barnevelder female

Barnevelder

Like the Welsummer, the Barnevelder is from Holland. Colour-wise, by far the most popular and arguably the most attractive is the double-laced version. The edge markings of the cockerel's feathers appear iridescent green in the light and its tail feathers are black, but on the hen bird, the lacing continues right into the tail. There has, unfortunately, been more emphasis placed on the number of eggs laid rather than on their colour in recent years and so, if it is the traditional dark brown-coloured egg that attracts, it will pay to be very careful when sourcing your stock.

As a bird for the back garden, it is docile and quickly becomes very tame indeed. In fact, according to *The Complete Encyclopedia of Chickens,* the bantam version is particularly known for being 'calm and affectionate birds' and 'both in a run or wandering about freely, they thrive'.

Croad Langshan

Further alternatives that may be usefully considered by the explorer in search of a coloured (rather than white) egg for breakfast include the Croad Langshan. It is a lovely heavy bird and a reasonable layer of eggs that vary between light brown and cream. It is quite hardy and therefore relatively easy to keep – as with all the chickens mentioned so far it soon settles into a back garden family environment. The breed is mainly seen with black plumage although a white variety also exists. Somewhat confusingly, there is also a separate somewhat lighter (in weight rather than colour) Langshan breed developed in Germany which will lay more eggs than the Croad (which was created in Britain by a certain Major Croad and was exported to Germany, where it was used to develop the German Langshan). Both breeds have a single comb, but the German breed has clean legs, whereas the Croad Langshan has sparsely feathered ones. German Langshans are also typically wine-glass in shape while the old-fashioned Croad is more rounded in conformation (for a photo showing a perfect trio of German Lanshans, *see* page 29).

New Hampshire Red

Developed as a utility bird in America, there is in the make-up of the New Hampshire Red the genes of the Rhode Island Red, which it resembles in size and stature – although its coloration is more ginger than red. Like the Rhode Island, it lays quite well (the eggs are tinted to light brown) and makes an equally good addition to the back garden. Although all my chosen breeds ought to be relatively easy to access, sourcing New Hampshire Red stock in either large fowl or bantam type might not be so simple. There is, however, a breed club in Britain which has over thirty members so, if you join, it might just be possible to acquire some birds from one of them – especially if you are prepared to wait until any surplus stock becomes available.

New Hampshire Red male

Plymouth Rock

You will more than likely see a preponderance of barred Plymouth Rocks rather than any other colour at a show. However, black, buff, Columbian and white also exist. In the barred coloration, it will be noticed that the hens are darker than the cocks, which is all to do with genetics. Plymouth Rocks are good layers of tinted eggs and are known for being friendly birds and so, as well as being practical, make good pets. The breed's placid nature makes them ideal for a relatively confined area as, although like most breeds they appreciate being given free range, they are almost as happy when penned. The hens, in common with most other soft-feathered, heavy breeds, are excellent sitters when broody and even better mothers; they are therefore one of the better options for anyone new to chicken-keeping who wishes to hatch chicks naturally.

PRETTY

Pretty can, of course, include practical, but for the purposes of this section, I shall include those breeds that are more recognised as exhibition birds rather than for their egg-laying abilities. There is a great deal of pleasure to be had in showing birds, but even if you don't, any of the following make an attractive addition to the garden.

Sebright

Can there be anything more pristine and beautifully marked than either the silver or gold-laced Sebright bantam? Can there be any other diminutive breed that struts about the garden and stands in the show pen so proudly? I would suggest not. Provided that you are aware of their flying abilities, I don't think there is an exhibition-type bird more suited for the back garden and more able to give their owners pleasure. The only possible downside might be that breeding future stock to show standard may be difficult, owing to the importance of the correct lacing. However, obtaining good quality birds at the start will make the development of a successful strain far more likely. The point must also be made that the fertility of some strains of Sebright has suffered as a result of inbreeding and also the importance placed on hen tail feathering in the males. Colour-wise, it might also nowadays be possible to find yellow white-laced and lemon black-laced variations but I would venture to suggest that they do not compare to the pristine colours of the originals.

Silver Sebright female

Sir John Sebright developed both the gold and the silver variety at the beginning of the nineteenth century – making it one of the older British varieties of true bantam. Like the Rosecomb bantam, whilst the Sebright undoubtedly makes an ideal choice for the small garden, they can only ever be expected to lay relatively few creamy-white eggs (perhaps sixty to eighty in a season). Their huge characters do, though, more than make up for a lack of breakfast and their tiny stature – they are, arguably, the best possible proof that very few gardens, no matter how small, cannot accommodate at least a trio of chickens of one sort or another.

Modern Game

A bit like the advertising executives of the 'Marmite' product will have you believe, you will either love or hate Modern Game – I don't think there can be any half measures! Long-legged, haughty and proud is a description perhaps more suited to a cat-walk model than to a chicken, but the comparison is totally appropriate. This is a hard-feathered breed that originated in England and was developed sometime in the late nineteenth century from fighting game fowl stock. After the abolition of cock fighting in 1849, and with the increasing popularity of showing, judges started looking for taller birds with a shorter hackle and smaller tails. They were originally known as 'Exhibition Modern Game' but the word 'exhibition' was later dropped.

Examples of the many colour variations are best seen at some of the larger poultry shows and include black, black–red, blue, golden duckwing, pile, silver duckwing, wheaten and white. Being light in build and therefore a little flighty, they will not generally become as docile as some other breeds (but I have, over the years, seen some very tame examples), and if you're looking for a decent-sized egg for your breakfast, search elsewhere! As with all breeds they do lay eggs but they tend to do so more in the spring and early summer rather than year-round. While both large fowl and bantam varieties exist, the large type arguably had its heyday in the early part of the last century, and it is the bantams that are nowadays more popular – a plus point when it comes to sourcing a small, pretty bird to enjoy in a back garden environment.

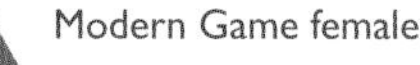

Modern Game female

Cochin

I was at first unsure whether to include this particular breed in this section, or to wait a while and mention it as being suitable as a 'pet'. It is perfect for either definition but, when I considered the fact that, when the Cochin was first introduced to Britain in Victorian times (*see* also page 160), it was mainly because of its exhibition qualities that it caused such a stir amongst poultry fanciers, I eventually thought it best suited here. In fact, Chris Graham, in his book *Choosing and Keeping Chickens* (Hamlyn, 2006) stated that 'many experts believe it to be the breed that spawned the whole poultry-keeping and showing hobby that we have today.'

The Cochin is a big, ponderous bird which can be found in a variety of colours: black, blue, buff, cuckoo, partridge (which has lacing on the neck, wing and back feathers) and white. Despite having feathered legs, they are generally easy to care for, but, if you wish to show them, it is important that they are not kept in such a way that the leg feathers will become broken, dirty or otherwise damaged. Fortunately, the breed is not all that active and certainly is not one that spends its time scratching, dusting and otherwise making a nuisance of itself in the garden.

As to the numbers of eggs it is likely to lay, it is, in common with most exhibition breeds, not all that brilliant a layer; what it does produce are light brown in colour. On the plus side, the hens are superb sitters and mothers so you should, almost always, have at least one of your pen of birds that will go broody and can be used to sit the eggs of the others.

Cochin male

Brahma

Another feather-legged big (or perhaps I should say 'huge') bird is the Brahma – it most definitely makes a statement both in the garden and on the show bench! There is a school of thought which suggests that there is some Cochin in its origins. The breed arrived in Britain (where it was at first known as either the 'Brahma Poutra' or 'Brahmaputra') at roughly the same time as the Cochin (the mid-nineteenth century) and quickly became almost as popular. Their loose fluffy feathering make an already large bird seem truly enormous but despite this, children seem not to be at all fazed by the breed – with the result that they are huge favourites with both children and adults alike. They are very easy to tame, and will readily follow you around the garden, albeit with an ambling gait – which is due to their feathered legs and feet.

Brahma male

Victorian, Edwardian and subsequent back garden chicken-keepers have always found the breed to be docile and friendly, and although they have never been a breed commonly seen, they have always been somewhere in the background and have always had a devoted, if somewhat minority, following. Maybe it is their size which puts people off, but, despite being a big bird, they don't require any more feeding than the average chicken, and their slow, casual demeanour means that they actually require less space than some other breeds.

For a heavy breed, the hens lay a surprising number of eggs, which are light brown in colour. The comb is known as being 'triple' or 'pea', and feathering can be blue partridge, buff, Columbian, blue Columbian, dark, gold, light or white. Finally, it is interesting to note that Brahmas are considered to be slow to mature in comparison with many other breeds.

Frizzles

Frizzles are a most unusual exhibition breed having, as they do, a feather structure which curls towards the bird's head rather than, as with most other breeds, lying neatly along the body in the direction of the tail. In

Frizzle male

some countries, the word 'frizzle' simply denotes a feather type but in the UK it is a definitely recognised breed with its own standards. Despite being an exhibition breed, the Frizzle lays reasonably well and, although it may seem that their unlikely feather formation might leave them unable to cope with particularly extreme weather conditions (imagine a house roof with its tiles up-ended rather than laid and securely fixed to under-pinning batons!), they are a quite hardy bird which is certainly well able to cope with life in the back garden. Because of their appearance, children seem fascinated by them (as they do with Silkies – *see* page 73) and so they make good family pets as well as extremely interesting show birds.

Breeding for show might, however, be a different matter because, as a result of their genetics, chicks from the same hatch may include individuals that will eventually display different (and unwanted) feather types. If you were to breed from the same strain year on year there is also a danger that you will end up with birds that have weak or sparse feathering. In order to avoid this, experts generally recommend that, periodically, a fresh, well-frizzed male is used as a mate for your hens. Colour-wise, there are many variations, but the comb and wattles are always bright red. If you have no intention of showing, it might be quite fun to build up a small flock of mixed colours. The eggs are cream or tinted.

Japanese

The Japanese breed is another example of a true bantam, and perhaps its most noticeable features are the very distinctive, long upright tail carriage of the cock bird and the short legs of both sexes – so short in fact that they sometimes cannot be seen as they are covered by the thigh feathers. Rarely does one ever have any impression of them being in any other position other than sitting (although they are most often actually standing) majestically, and it might therefore be concluded that, rather like a supercilious Persian cat, there is nothing in the breed that will interest the back garden chicken-keeper. In fact, nothing could be further from the truth and they are an exhibition breed well worthy of consideration.

Originally, it would have been unusual to see anything but the feather type observed on most ordinary chicken breeds, but since the 1960s, two other feather variations have become increasingly popular amongst Japanese bantam show enthusiasts. The first is the frizzle type, whereby each feather should curl towards the head of the bird, while the second is similar to the stranded feathering of the Silkie breed (*see* page 73). The comb of all types should be single and evenly serrated with four or five points. There are many colours accepted within the standard, but it is the black-tailed white which is the most often seen.

Sadly, successful breeding is not always easy because mating and subsequent fertility can be difficult; in addition, the breed can also occasionally suffer from a genetic condition that arrests the development of the chick while still in the egg. Despite these potential difficulties, these pretty, very ornamental bantams are well worth considering by anyone looking for something a little unusual.

▲ Japanese male – their natural stance makes them look as if they're sitting rather than standing!

The Best of the Rest

If new to chicken-keeping, it is perhaps best to avoid any of the breeds which, because of feathered feet, complicated 'top-knots' or a flighty disposition, might require specialist knowledge and more complicated housing. However, in addition to the breeds already mentioned above, here are a few more you might like to consider as potential exhibition birds – some might be slightly more difficult to look after, but they are all, most definitely, extremely 'pretty' to look at.

Hamburg

Although the Hamburg breed came from northern Europe, both spangled and black Hamburgs were bred as large fowl for over three hundred years in northern England, where they were alternatively known as 'Pheasants' or 'Mooneys'. Classified as a light breed and soft-

A show champion Silver Spangled Hamburg male

feathered, the breed is, like the Sebright, very smart and elegant-looking. Hamburgs should have a rose comb (with that all-important 'leader') and white ear lobes. It lays a reasonable number of white eggs each year and would undoubtedly make a good bird for the novice owing to its ease of keeping. It must, however, be mentioned that, like a good many other light-coloured birds, the silver pencilled variety in particular may become discoloured or 'brassy' if exposed to too much sun – the feathers would moult out, but any suspicion of 'brassiness' would put paid to the current season's show prospects.

Rosecomb

A 'true' bantam, the Rosecomb is thought by many to be the oldest bantam breed in the UK and its existence was noted by writers in the latter part of the fifteenth century. The breed is full of character and, being small, would make an ideal exhibition/pet bird in situations where space is limited and where they might have to be confined to a moveable house and run. They are very pretty, proud birds and, while black is probably the most popular colour, other variations include barred, brown–red, buff, creole, exchequer, lemon blue, mottled, silver duckwing, wheaten and white.

Despite being very easy to keep on a day-to-day basis, Rosecombs can be difficult to breed because some strains are subject to male infertility; therefore, it will pay to source stock carefully and to try and gain some reassurance from the breeder that their birds have not been subject to inbreeding in recent generations – a practice known to encourage hereditary faults.

Poland

In view of the earlier comments regarding not running before you can walk, I include the Poland (sometimes erroneously known as Polish) with some trepidation. It is not one of the easiest to keep, but then again it is certainly not the most difficult and it is such a striking exhibition bird that it would be a shame to miss it out completely. Despite any potential difficulties, the breed does have many devoted followers – mainly due to its 'top-knot' or crest (which occurs because the feathers that create it emerge from a protuberance at the top of the skull). The colour variations are many and varied and, for its size, it lays a surprising number of white eggs during the season.

Malay – looks-wise, the 'hard nut' of chickens!

Malay

Should anyone fancy a brute of a bird which by no stretch of the imagination can be termed 'pretty', but is nevertheless becoming increasingly popular among the showing fraternity, then what about the Malay? The breed is described in some standards as having a 'cruel and morose expression', but looks can be deceiving and, compared to several other hard-feathered breeds, it has a reasonably placid nature: while the males might be aggressive towards each other, both cocks and hens are quite happy around people. Children, especially young teenagers, love it (as they do the Shamo or Ko-Shamo, which, with its piercing eyes, is a breed equally mean-looking). The Malay is, however, quite an active breed and requires plenty of space. As far as showing is concerned, the size and shape of the bird is more important than its colour, which can be black, black–red, pyle, spangled or white.

PETS...

As pets for children, bantams cannot be bettered. Their character endears them to most people, but they can be particularly rewarding for youngsters, giving them a practical knowledge of biology, livestock management, and even day-to-day discipline, as they begin to realise that bantam-keeping is a seven-day-a-week undertaking. Some are, of course, more suited than others and it will probably be sensible, especially if you have toddlers, to opt for one of the heavier, generally friendlier breeds of bantams whose solid and staid nature is more likely to appeal. It is, however, important to remember that any child's natural reaction is to try to pick things up, so it is essential that they are taught how to hold both bantams and chickens from the very outset.

Orpington

Both the large fowl and bantam Orpington make excellent pets but, perfect though the large fowl is as a beginner's bird, the bantam is in a league of its own when it comes to being the absolute children's pet. As I remarked in *Choosing and Raising Chickens* (David & Charles, 2009), 'its friendly, docile nature makes either the large fowl or the bantam an excellent choice for the first-time chicken-keeper'.

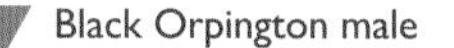
Black Orpington male

Cuddly and attractive to look at, as well as being very easy to tame, there is the added bonus that although the cockerels will crow occasionally, they do not feel compelled to do so all day long (as is the case with some other breeds). Their conformation means that they have little, if any, inclination towards flying or rushing around and so it is possible to contain them in a wire-netting surround of no more than about one metre (provided that by doing so they are not placed in any danger from predators or wandering dogs). Also of benefit when considering a children's pet is the fact that they will very easily go broody and make excellent and reliable mothers, so it should be possible to collect a clutch of eggs and place the broody hen in a suitable, safe and secure place in order to allow young children the opportunity to see the whole incubation, hatching and rearing sequence from beginning to end. Standard colour variations include black, blue, buff and white. Their egg coloration is light brown but whereas at one time it was an excellent producer of eggs, some modern strains will not necessarily lay that many. This is thought by some to be the result of exhibition breeders mating birds for size and an abundance of fluffy feathers around their vent area rather than for their egg-laying abilities.

Gentle and happy enough with those of their own kind, beware of mixing them with a pen of assorted breeds as they, like Brahmas, can be subject to bullying (*see* 'Mixing Breeds' on page 52).

Dutch

A pair of Dutch bantams

If space is limited, and you don't like the Sebright or Rosecomb (shame on you – how can you not?) you might just possibly take a shine to the Dutch breed of 'true' bantam which requires very little in the way of suitable accommodation. It is, like the two previously mentioned breeds (and several more besides), clean, sleek-feathered and pristine-looking. It was first standardised by the Dutch Poultry Club of Holland in 1906. Bred from several regional sources, the breed very quickly became popular throughout its home country and, despite only being brought into Britain in the early 1970s, it very quickly established a reputation for being an ideal breed to be kept in a small area or a confined space – although, like most breeds, the more space you can give them, the better they will fare.

The colour variations are quite amazing – there are more than twenty recognised as standard – and this, coupled with the fact that they are a hardy breed, which seeks out and thrives on loving care and attention, makes them the most perfect of pets. You need not miss out on your breakfast either because Dutch bantams have been known to lay over one hundred and sixty eggs in a season. Having said that, their light-coloured eggs do tend to be quite small and you might be better making omelettes rather than relying on a single one as a boiled egg! Surprisingly, for such small birds, the females make excellent and reliable mothers. There can be few breeds better suited to the back garden environment or many that are more friendly – they can, however, fly quite well so it will pay to make sure that they don't go exploring the neighbour's garden as well as your own!

Silkie

Unlike some of the other breeds I've suggested in this section, I do not know of anyone that isn't taken by the charm and unusual appearance of the Silkie breed: they seem to appeal to children and adults alike. Their quiet, gentle character makes them a perfect pet, suitable for almost everyone. They rarely become alarmed and, therefore, have little or no inclination to flutter about madly at the slightest disturbance. Their most obvious feature is the unusual feathering, which looks more like strands of fur than feather. The breed also has a small crest or 'top-knot' which looks like a little pom-pom on the females; both male and female have feathered legs.

The crest and legs can get a little soiled in wet, muddy conditions and so, where possible, it might be best if they can be kept in a covered house and run during any prolonged spells of inclement weather. Here I must tell a personal tale. When my daughter was small, she had a trio of very pretty Silkies; unfortunately for them we also had a small swimming pool. Experience now tells me that Silkies and water do not mix and, most probably due to the absorbency of their feathers, they are quite prone to drowning. My daughter eventually ended up with a totally new trio of birds due to the fact that, over a period of time, they'd all fallen foul of the pool (fortunately whilst she was at school) and had, unbeknownst to her, been replaced by three separate mad-dash trips to the original breeder – after which time, I built a bigger pen and forbade their free range of the garden! They do (when not drowning) lay a creamy-coloured egg. As to the bird colours themselves, they include black, blue, gold, partridge and white – the white is, however, a little boring and I personally would choose one of the other colours.

The comb of a Silkie is mulberry-shaped and, quite unusually, purplish in colour rather than red. It also has a crest – as can be quite clearly seen on this particular cock bird

Pekin

I wasted so much time when, at one point in my chicken-keeping life, I decided to keep a pen of Pekins – I also wasted much of my lunchtime sandwich because, as I sat out on the lawn in the summer, I would very quickly be joined by our pen of free-range Pekins which could charm whatever they wanted from whomever they wanted!

Pekin bantams are docile, friendly birds that lay reasonably well, come in a variety of different, very interesting colours (the lavender is particularly attractive) and, despite their delicate puff-ball look, are quite hardy (British standards stipulate that the Pekin should be single combed, and '…circular in shape when viewed from above, [the] whole outline rounded…'). The hens make excellent mothers, rarely abandoning the nest before the hatching period is over and being naturally careful when scratching about around their newly hatched offspring. Conversely, the fact that they are so heavily feathered about the legs and feet makes them less suitable for heavy clay soil and low-lying areas that are prone to wetness.

There is sometimes some confusion between the bantam type of Cochin, which has a large poultry counterpart, and the Pekin, which is classified as a 'true' bantam by the Poultry Club of Great Britain. Both were imported from the Far East in the mid-nineteenth century and some Pekin enthusiasts believe that the first birds were stolen from the private collection of the Emperor of China some time towards the end of the Opium Wars around 1860. Others believe that today's Pekin is a derivative of the Cochin – whatever the truth, they make great pets and will follow you round the garden whilst all the time appearing to be telling you their life story – in reality though, they are probably just after your summer sandwich!

Silkie crosses make good broodies

Some chicken and bantam breeders keep a pen of Silkie crosses to use purely as sitting hens for clutches of eggs coming from their main interest breeds which are not, for a variety of reasons, likely either to go broody themselves, or make poor mothers when they do. The fact that a cross is preferred is because, while they have all the superb mothering tendencies of the Silkie, due to the cross not having fine, cotton-like feathering, they are considered less likely to strangle the brooding chicks with their luxuriant feathers – as might possibly be the case when using pure-bred birds of the same ilk.

A Mottled Pekin female

Even the female Faverolles have a quite pronounced and obvious 'beard' and side muffs

Faverolles

Just as quirky and charismatic as the Pekin bantam mentioned above, Faverolles have quite considerable charm and personality – they also have, somewhat unusually, five toes (like the Dorking breed) rather than the generally accepted four. Both the large fowl and bantams also have lightly feathered legs, a beard and side muffs, making them a very attractive addition to any back garden. The muffs probably add most to their pet appeal due to the fact that, as the birds look quizzically at you, the muffs look like ears, and therefore the birds appear to be listening to your every word (what do you mean, you don't talk to your chickens – everyone else does!).

Colour variations permitted in the British breed standards include black, blue-laced, buff, cuckoo, ermine, salmon and white. The blue-laced is particularly attractive and is dark blue with each feather edged with a darker shade of the same colour, but it is arguably the salmon coloration that is most commonly seen.

Although they will never lay the greatest quantity of (creamy-coloured) eggs, such a fact is perhaps not surprising, as, when the large fowl version was originally bred in France, it was classed as being primarily a table bird and any amount of eggs the female laid over and above those required for hatching was a bonus. Present-day Faverolles enthusiasts all seem to agree on the fact that the breed is quiet, calm, docile and easy to handle – surely the perfect attributes for any pet? They do, however, also say the breed is prone to being bullied and so it is probably not a good idea to include the odd individual in a mixed flock (*see* page 52).

Araucana

Although Araucanas, being light of stature and therefore potentially flighty, might not be everyone's first thought when it comes to choosing chickens suitable as pets, they are, nevertheless, reasonably easy to keep and soon settle into a back garden situation. However, the main reason for their inclusion in this particular list is because their eggs are such a talking point. Most famous as producers of blue-shelled eggs, the breed also has an interesting and ancient history. Named after the Arauca Indians of Chile, the original breed was rumpless, but birds which from their description can only have been Araucanas were commonly seen around the Mediterranean from around the middle of the sixteenth century. It is probable that traders exporting Chilean nitrates introduced the birds to Europe, where it is nowadays more common to see birds that possess a tail or rump. So, not only is there the opportunity to pick up blue eggs, but also the opportunity of owning a breed with an ancient history.

Araucanas have pea combs, red ear lobes, ear tufts and, when *not* of the rumpless type, a head crest and muffs. There are several plumage variations, the most popular of which is probably lavender – others include black, cuckoo, fawn, silver-blue partridge, wheaten, yellow partridge and white. With plenty of colour types to choose from, the hens being reasonably good layers of most distinctive blue eggs (perhaps anywhere between one hundred and fifty and two hundred in a season), and of an ancient lineage – what more do you need as a talking point when the neighbours next come round for supper?

An Araucana female – white though this particular specimen may be, she will lay bluish-coloured eggs

The Best of the Rest

Before mentioning two or three further possibilities, it must be reiterated that there are many breeds that will make suitable pets and, in reality, it has more to do with the attitude of their owner than the choice of a particular breed. One of the best stockpersons I have ever known, takes time to sit with all his animals and poultry, and moves about them in a slow, gentle manner; he could, I am sure, calm a crocodile and feed a tiger tasty tit-bits without any mishap!

Serama

Well known as being possibly the smallest bantam in the world, for anyone who has the most limited of space (too small even for a couple of Sebrights or Rosecombs…I don't believe it!) and yet wants to enjoy a few chicken pets in the garden, the Serama might just be the absolute answer. The fact that a trio can be housed in nothing more than a decent-sized rabbit hutch and require only a little simple day-to-day care and attention makes them well worth considering. As well as having limited accommodation requirements, their temperament makes them ideal candidates as pure pets.

Seramas have a single comb and come in almost any colour variety. Interestingly, they do not necessarily breed true and parents of one colour may produce chicks of several different variations. Do they lay eggs? Well, of course they do – but not many: fewer than one hundred in a season.

Jersey Giant

From the sublime to the ridiculous! If the Serama is the smallest type of domestic poultry, then the Jersey Giant must be one of the largest (there is, I suppose, a clue in the name!). If space allows, this breed of American origin is, according to many, a calm, gentle giant which, unusually for a very heavy bird, is capable of laying almost two hundred brown to light-brown eggs in a season. Like some of the other particularly large breeds (Brahmas, for instance) it is slow to mature, but there is no problem in that. There are three colours: black, blue and white (the blue actually has a laced pattern to its feathers) and, although they may be difficult to source, they most definitely make good pets.

Whether I shall ever achieve it, it is my ambition to one day own a pen of these birds (and Brahmas…and genuine Croad Langshans…and almost every kind of Wyandotte…and a multitude of others besides…ah well, we can all only dream!).

Jersey Giant female

Scots Dumpy

Somewhere in between the two extremes of Serama and Jersey Giant comes the Scots Dumpy (infrequently spelt 'Dumpie'). A native and ancient breed (its ancestors were believed to have been brought to Scotland by Phoenician traders well before the Roman invasion of Britain), the traditional reason given for its noticeably short legs and distinctive waddling gait is that, in Scotland, where the crofts were surrounded by wild countryside in which predators were rife, its short legs and heavy body encouraged it to stay at home! Whatever the truth, it is a calm unexcitable breed which certainly makes it well worth considering for the back garden situation.... A loss of genetic material meant that, until relatively recently, colour variations were limited to either black or cuckoo, but it is nowadays possible to find brown, gold, silver and white.

HYBRIDS AND 'RESCUE' HENS

So far we have talked exclusively of pure-breed chickens, but what of those chicken-keepers who prefer hybrids, or to rescue laying birds after their commercially useful life is over? The British Hen Welfare Trust is a national charity that re-homes such hens, educates the public about how they can make a difference to hen welfare, and also encourages support for the British egg industry. Its ultimate aim is to see consumers and food manufacturers buying only UK produced free-range eggs, resulting in a strong British egg industry where all commercial laying hens enjoy a good quality life. In the meantime, they find homes for thousands of commercial laying hens otherwise destined for slaughter and are actively supported by celebrity patrons such as Jamie Oliver, Anthony Worrall Thompson, Amanda Holden, Pam Ayres and Jimmy Doherty.

According to Valerie Hardy, a writer on all things rural and a passionate chicken-keeper: 'hybrids make wonderful pets and ours have

Not all animals are equal

As George Orwell famously remarked in *Animal Farm*: 'all animals are equal, but some are more equal than others'. Some of the hybrid breeds are used in commercial production as 'intensive', 'battery' or 'cage' hens. While the UK has recently made moves towards enhancing a bird's quality of life by allowing more space per bird and including scratch material, perches and actual nest boxes rather than rolling trays ('enhanced' cages), other countries have yet to follow. In 2010, Poland proposed a delay to the EU ban on conventional battery cages until 2017 (in effect, this would allow Member States to continue keeping hens in the most basic of cage conditions until then).

Young 'Bluebelle' hybrids – attractive and good layers

become quite used to my nine-year-old granddaughter picking them up and carrying them all over the place.' One particular producer of hybrids opines that they are ideal for beginners to the fancy and that they 'tend to be more disease resistant than pure breed chickens. Furthermore, they are usually available all year round and are cheaper than pure breeds.'

However, Francine Raymond, author of *Keeping A Few Hens in Your Garden* (Kitchen Garden, 1998) wrote in a recent article that, in her opinion, while 'pure breeds tend to have a life expectancy of eight to ten years, hybrids have a shorter lifespan (three or four years)'. In doing so, she echoes the thoughts of several online 'bloggers' who also maintain that hybrids lay well but burn out more quickly than do pure-breeds. In the same article, Francine goes on to state that keen gardeners need to be aware of the possibility that 'hybrids will eat everything if you let them roam around'. Why she should think that hybrids are potentially capable of doing more damage to a garden than pure-breeds, I've no idea. In fact, I agree more with Charlotte Popescu, author of *Chicken Runs and Vegetable Plots* (Cavalier, 2009) who comments, 'You will soon discover which vegetables and plants you must protect from your hens and the ones that your hens will happily avoid.'

What is a hybrid?

Basically, a hybrid is the result of crossing two or more breeds. The offspring will not, however, breed true and reproduce chicks in their own likeness and it is therefore necessary to go back to the original parent stock and breed them again for more of the same. Hybrids are generally commercially fashioned and are often given weird and wonderful names by their producers – no matter what title they are given, they are not, in any shape or form, actual 'breeds', which is, to my mind, a very important point to remember. They are, though, very efficient egg layers and if that is what you are after, then you can do no better than order half a dozen or so. You could even build up a little flock consisting of different varieties of hybrids, but if you do, try to buy them all at the same time and all of the same age – usually 'point-of-lay'. Some people consider hybrids to be a bit on the boring side as far as looks are concerned, but there are, in fact, some quite attractively coloured hybrids on the market.

You could, of course, keep an ever-changing flock of hybrid birds by periodically introducing totally unrelated cock birds but if you do, the resultant chicks will be cross-breeds or 'mongrels' rather than hybrids and the problem with cross-breeds is that any desirable points that might have been so obvious in the original stock may well be diluted in subsequent generations.

EXPLAINING 'HARD' AND 'SOFT' FEATHER ...

Most breeds are 'soft-feathered', except for those which are not! The overall thinking is that soft-feathered breeds are those where the plumage is quite loose and fluffy. Examples include all of the breeds mentioned in the 'practical' section earlier in this chapter, all bar Modern Game and Malay in 'pretty', and all those discussed in 'pets'. Being soft-feathered generally means that the actual body shape is somewhat disguised by the feathering, whereas those breeds which are considered to be 'hard-feathered' are generally of the game fowl type – the feathers of which very tightly follow the body shape and do not distract the viewer's eye from the overall frame and structure.

Bantam Old English Game male

Should you be considering showing your stock, it is accepted that the plumage of soft-feathered breeds will be washed before being exhibited, while hard-feathered examples will not (if you did, the feathers would lose some of their firmness and definition, making it more difficult for the judge to assess their all-important body profile – of which more in Chapter 8).

... AND 'HEAVY' AND 'LIGHT'

Anconas, Leghorns and Minorcas, indeed all Mediterranean breeds, are classified as being 'light'. 'Heavy' breeds include the Jersey Giant, Rhode Island Red and Wyandottes. The former are, more often than not, exceptional egg layers but, on the downside, tend to be a little flighty in their character. The latter are quite often dual-purpose breeds and are, with notably few exceptions, calm and placid – which more than makes up for the fact that they tend to lay fewer eggs throughout the year (their easy domesticity is the main reason I have included so many examples within this particular chapter).

For a full definition of breeds which are soft or hard-feathered; heavy or light, one need only refer to the current book of *British Poultry Standards*, the last of which was published in 2008 (it is generally revised and reprinted every ten years).

5

PREPARATION, PURCHASE AND PRACTICE

There is a bewildering array of feeders and drinkers – at a variety of prices!

Once the reasons why you want to keep chickens have been established, the best place to keep them resolved, and the decision made as to what chickens will best suit your particular lifestyle, there's not really a great deal left to do except to go out and buy them. Or is there? Well, yes, actually there is because, while the subject of suitable floor and nest box litter has been mentioned in Chapter 3, there is still the question of getting ready for the great day by acquiring suitable feeders and drinkers, and feed appropriate for the age of birds you are intending to buy and also making a few last-minute preparations.

FEEDERS AND DRINKERS

Feeders and drinkers will obviously be required before any birds are purchased and a glance through any catalogue or website will reveal a huge and often confusing choice. It might therefore pay to seek out an experienced chicken fancier (perhaps at a show, where there will also be trade stands selling equipment and you can see for yourself exactly what

Not content with his terrier pup, a little rotund pony with which he followed hounds in the company of George the groom, Young James decided, at the age of nearly ten he wanted some 'fine-looking hens'. His mother Esther, ever one to indulge the child, gave instructions to the gardener to build a large coop and enclosure in the corner of the walled garden and to equip it with all that was necessary. Come the morning of Young James' birthday, not long after breakfast, we all went for a supposedly aimless ramble in the gardens where we suddenly discovered, very much to the boy's delight, a pen of the finest-looking birds imaginable. An excited rummage for eggs ensued but, disappointedly, there was, as yet, nothing in the pristine nesting box…

Taken from *Visiting Home* by J. E. Marriat-Ferguson, published privately in 1905

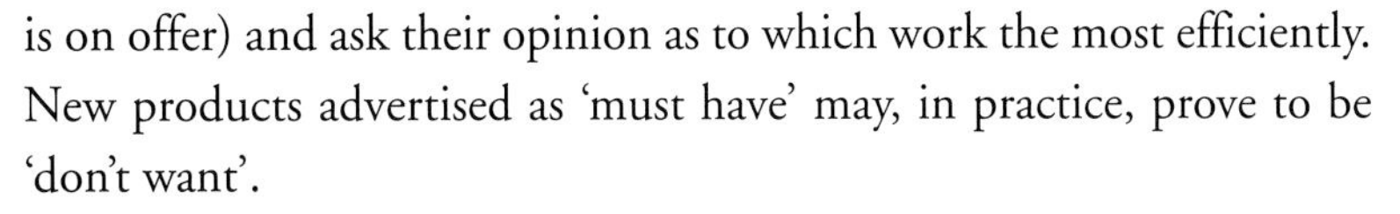

is on offer) and ask their opinion as to which work the most efficiently. New products advertised as 'must have' may, in practice, prove to be 'don't want'.

The subject of what type of feeder to use for garden chickens can be a tricky one. If your birds are in a combined house and run, it is a simple matter to keep the food dry in an open-topped trough; however, the food will soon become covered by floor litter or grass strands as the chickens scratch and dust. The generally low height of a house and run arrangement will probably preclude a hanging feed hopper which, in other situations, would go some way towards alleviating this problem. Chickens given full or partial free range of the garden are perhaps a little easier to cater for, but even so, there is a difficulty in finding a truly weatherproof feeder that will not attract vermin and wild birds – the latter can take a surprising amount of expensive feed in a very short time (*see also* 'Back garden pests' in the next chapter).

I have recently discovered what may well prove to be the answer, in the form of a sturdy 5kg capacity treadle feeder made from galvanised steel. This particular feeder features a close-fitting lid and handle, and an adjustable treadle plate for birds of different weights. The feed trough is opened when a chicken steps on the platform, and is closed as soon as it gets off. Most importantly, lighter animals such as rats, mice and wild birds will not be able to access the feed, thus lessening the risk of encouraging vermin, and saving on overall feed costs owing to there being very little wastage. Although sometimes perceived as being somewhat simple creatures, chickens soon learn to operate the system, which also has the incidental advantage of occupying birds longer by its operation than when they simply have to peck away at an open-topped trough or similar.

This particular feeder features a close-fitting lid and handle, and an adjustable treadle plate *(Photo courtesy of David Bland)*

Chick drinkers

To my mind, galvanised drinkers are best – this particular design has been around for a very long time

If you only see your birds twice daily, make sure that their water container is large enough to ensure that they do not run out on even the hottest day – water intake can increase by as much as 25 per cent in hot weather. While drinkers for adult birds might be a reasonably straightforward issue, those for very young chicks need to be carefully considered. Newly hatched chickens are less likely to drown if the right type of drinker is used. There are many chick drinkers on the market, most of them of the plastic twist-lock type, but despite all the recent innovations made to cash in on the current trend for chicken-keeping, I personally don't think you can beat the tried and trusted old-fashioned inverted jam-jar drinker. All such drinkers work on the vacuum principle, but simple is most definitely best in this instance – do not be tempted into purchasing the large type with integral handles because the 'trough' part is still too deep for very young chicks. Choose instead the ones that cost about five pounds where a standard jam jar fits into the galvanised base; even so, they are not fool-proof and, should you be raising the chicks of very small breeds such as Serama bantams it will pay to cut a small piece of very thin hose that fits into the base of the trough and makes the available water even more shallow. Experienced chicken-keepers may suggest using small clean pebbles to achieve the same purpose, but doing so is more time consuming when it comes to cleaning out and replenishing the drinker. The one disadvantage to such drinkers is that they can be knocked over by an over-enthusiastic broody hen as she scratches in the floor litter or grass run, but even so, they are, in my opinion, by far and away the best. To ensure that they remain level on a bed of shavings, place the drinker on a small square of hardboard (rough side up) – this also helps keep the drinker free of shavings and other litter debris.

It should go without saying that drinkers and feeders must not only be thoroughly cleaned from time to time, but also be moved periodically in order to prevent a build-up of potential disease problems as a result of a concentration of mud, faeces and/or wet litter around their base.

FEEDING

It makes absolute sense to feed only the best food in order to get the best results in terms of overall health, egg-laying, fertility and show condition. Commercially produced balanced feeds are available from local agricultural suppliers and pet shops (although the latter may well prove a little more expensive) and come in pelleted, crumbed or mash forms. They all contain essential vitamins and minerals and are available for all types and ages of poultry.

Compound feeds are often fed ad-lib – and the treadle-type feeder described earlier is perfect for this – but, whilst it is important that your birds are given a top quality feed that is high in proteins and fats, unlimited access could cause them to become obese. As well as affecting a bird's general health, obesity can cause problems when it comes to egg-laying and fertility. For this reason, traditionalists often prefer to feed a certain amount of pellets or mash in the morning and then give a late-afternoon/early-evening feed of cereals. Fresh feed is paramount to keeping healthy chickens: it is very important not to place fresh food on any stale pellets or mash that may remain in the bottom of the feed hopper. It might well look as though there is still food left in the feeder, but can you be sure that it is not by now soiled and sour, or, in the case of cereals, just unpalatable husks rather than wholesome grains?

As to 'treats', despite what is sometimes advocated on several chicken-keeping 'blogs', I would steer clear. I have seen raisins (which can ferment in the crop) and even chunks of pineapple (which, given frequently, will result in some very unpleasant cases of diarrhoea) recommended, but all are totally unnecessary as far as the diet of your birds is concerned. Sometimes a little 'tit-bit' might prove useful in

There is an easily available supply of manufactured feeds – all of which are carefully prepared to give optimum benefit to your birds

humanising certain individuals – in preparation for a show, for example (*see* Chapter 8) – but it must be of a sort that will not upset the digestive system and overall dietary equilibrium.

What to feed

Well, it's all pretty straightforward really – what you see written on the bag is what is in the bag! Thus it is that chick crumbs are suitable for chicks up to the age of four weeks or so and growers' pellets are designed to take your birds until early maturity (late teenage years in human terms). There follows layers' pellets (or mash), but don't be in too great a hurry to feed chickens with these too soon because it is a mistake to have your pullets laying before they have reached maturity (depending on the breed, generally somewhere around sixteen to twenty weeks). Should your stock be intended specifically for breeding the following year, they can be kept on a regular diet until perhaps just after Christmas, when breeders' pellets might quite profitably be introduced because these have additional ingredients that go a long way towards ensuring that the parents are in prime condition and the chicks are born healthy. The occasional clutch of eggs set under a broody just for interest and amusement will, however, hatch perfectly well if the parents have been fed on a layers' ration.

Cereals

The reason that most traditionalist chicken-keepers like to feed their birds grain in the late afternoon/early evening is because it remains in the crop longer than a compound feed and will keep the birds feeling full and warm overnight. It does most certainly make the owner feel better to know that their birds are tucked up warm and replete on a cold, snowy winter's night!

Basically, the cereals suitable for chickens are wheat and maize – if you buy a bag of mixed corn from your supplier, it will also probably contain barley but in my experience, barley is not particularly favoured by

With a little time and patience, chickens will, quite literally, 'be eating out of your hand'

chickens and they will leave it in preference to the other two types. Maize can be bought either 'whole' or 'split' (in certain parts of the country, it might also be known as 'kibbled'). The benefit of split or kibbled over whole maize is that it takes birds longer to forage for it and therefore keeps them amused over a greater period of time; the disadvantage is that it generally costs more! It is possible to start feeding youngsters a cereal grain from the age of about ten weeks but I wouldn't give them too much to start with because they are still growing and need the correct balance of protein and other nutrients which can only really be found in their growers' pellets or mash.

In any case, one needs to be sparing in how much grain one gives a chicken of any age. It has already been mentioned that too much maize will create unwanted fat, but apart from that very important factor, it is wasteful to feed more than your flock will eat in perhaps a half hour period – you wouldn't believe how often I've seen uneaten wheat and barley actually growing in a run, which not only encourages vermin but creates a need for a combine harvester!

Greenstuffs

If allowed access to fresh grass and an occasional foray around the garden, chickens will pick up plenty of naturally occurring greenstuffs. Additional greenstuffs in the shape of the outer leaves of lettuce from your salad or the unwanted 'tops' of vegetables will, however, never go amiss. Any keen vegetable gardener (and chicken-keeping and vegetable growing go hand-in-hand) should be able to find enough under-developed greens – and even forked out weeds such as dandelions, groundsel and chickweed – to give their birds essential greenery. I remember one particular year, when I had so many outdoor cherry tomatoes that they were over-ripening on the vine before I'd had chance to pick them, that I used to throw them straight from the plant into the chicken run and my birds would stand there ready to catch them almost mid-air, for all the world as if they were auditioning to be part of a circus juggling act! If space allows in the garden, it might even pay to plant a row or two of greens solely for the purpose of feeding to your birds during the winter months. As I've mentioned before in several magazine articles, here in France it's possible to buy *chou fourrager*, a large cabbage-like plant that is sown at intervals during April, May, June and July and can be harvested from September through until February. It is in common usage among French chicken-keepers – why not seek out a packet of seeds on your next trip over the Channel and grow it for your birds?

Chickens and bantams will appreciate almost any surplus greenstuffs taken from the vegetable garden (hanging them from a string or nail helps prevent the greens from becoming soiled)

Water

Chickens need a constant supply of good quality fresh water – I am not suggesting that it should be of the supermarket bought, dinner party given, sparkling variety, but its cleanliness is very important. Provided that clean water is always readily available, chickens are less likely to be tempted into drinking from other sources which could be contaminated, and will thus avoid the possibility of illness. If, as most people do, you give water out in the run, try and position the drinker in a shaded part and if, for whatever reason, it is necessary to supply drinking vessels inside the roosting area, make sure that the drinker does not become full of scratched floor litter. The water will therefore be inaccessible or, worse still, full of faeces as a result of birds perching overhead. If you are feeding mash (to either chicks or adults), keep the water some distance from the food and watch out for fouling from the mash stuck to the bird's beaks.

As to where, in a small, combined house and run it is best to position the drinker, I recently had this question from a reader of one of the magazines for which I write: 'Where is the best place to put the water drinker? We currently have both the feeder and drinker on the floor of the run, but each morning our girls have moved it and the water is empty. My husband filled the water this morning and when I went out in the afternoon their water had moved and was empty. I know fresh clean water is important but our girls are being slightly naughty. What shall we do for the best?'

My reply was as follows: 'Unless your birds are to be confined to the house for any length of time, it is best to do as you are and keep the water in the run. When you say your chickens "move" their drinkers, I presume you mean that they knock it over with their scratching – which would suggest that it is a light plastic fountain type. Personally, I would only ever use galvanised fountain drinkers as they are more robust and, being heavier, less likely to get knocked over. I would also stand it on a brick so that it doesn't become filled with scratched grass and other debris. To answer your immediate question though, if it is a small run

Some plastic drinkers are easily spilt or quickly become filled with floor litter

attached to the hen house, I would stand the drinker in the corner (that way they are less likely to knock it over) and secure it by placing two house bricks tight against the fountain "bowl". Are you sure that the current drinker is big enough for the number of birds you have? They could be tipping it over when it is empty.'

Grit

Chickens of all ages need grit. When the birds are young, provide chick grit for the first six weeks of life and then move them onto adult hard flint grit. Looking at a sample of normal-sized grit, it might seem a little large for your teenage birds, but in amongst it will undoubtedly be some that they can quite easily manage. For chicks, sprinkle a tiny amount on their food, but thereafter leave a small container nearby to enable them to help themselves when required. Providing grit at such an early age helps the birds to build up a good strong gizzard which will make them far more efficient at getting the best out of their food.

Despite what others might tell you, if you are feeding a balanced commercially manufactured feed, it should never be necessary to give oyster shell (which is not actually a grit anyway) as, by doing so, there is a chance that its addition will imbalance the calcium phosphorous ratio that has already been included in exact proportions in the ration (for reasons of production, hard flint grit is never included by any manufacturers in their foodstuffs). Provided that a balanced compound feed is being given, there should be no problems with eggshell quality or texture – which is the reason some people advocate the use of oyster shell.

PURCHASING BIRDS

You might have already bought the best house, the most efficient feeders and drinkers, built a superb open-air run and be aware of how much your hobby is likely to cost you in set-up charges and in ongoing enthusiasm. You might even have taken the most sensible decision of all, which is to join a local smallholding group or poultry club and maybe also taken out a subscription to one of the several magazines which offer help and advice to poultry enthusiasts. It must now be time to ask whether or not you are ready to carry out the most important (and most exciting) part of chicken-keeping – which is, of course, the buying of your stock!

Be sure to buy the best

First of all, it is important to find a reputable seller. A well-known breeder has a reputation to maintain. By talking to the breeder and explaining exactly what it is you want from your chickens and where you are going to keep them, it should be possible to ascertain that their stock is suitable for your particular purpose. If so, try to assess whether or not

Some shows even incorporate an 'advice corner' where you can learn more about chicken-keeping – including how to handle them correctly

the examples on offer have been well looked after – birds treated well are usually quite obvious as they will often (but it does depend on the breed) have a far more docile temperament and gentle demeanour as a result of regular handling and constant care and attention. If it's young stock you're looking at, ask whether it might be possible to see any adult direct relatives of those you are considering buying because this will give you a good indication of how yours will eventually look. There are many breed associations with excellent websites on which you will find the contact details of well-known and well-respected breeders.

While visiting poultry shows is always a good idea before buying as they provide the perfect opportunity to learn something about the various breeds and are always attended by like-minded individuals ready and able to answer any question you might have; *never, ever* be tempted into making an impulse buy should there be a sale of birds included. It is all too easy to visit such a place and buy chickens on a whim – irrespective of whether or not you have anywhere suitable at home in which to keep the poor unfortunate creatures. Instead, only buy when

Characteristics of layers

There is, fortunately, one outward indication of good laying qualities of pretty general application to all breeds, and even to mongrel fowls, which should be borne in mind when it is necessary to purchase fowls about whose antecedents nothing is known for certain. This is what is generally termed the V-shape, or wedge-shape, formation of body. Verbally, it means that there should be a distinct downward gradient from the breast to the end of the pelvis (the bony cavity which forms the lower part of the belly) against a pretty good length of backline, when the bird is viewed sideways; viewed from behind there should be a corresponding widening out (like the top of an imaginary V), from the bottom of the ventral region. These characteristics are most easily noticed in the lighter breeds, such as the Leghorns and Anconas, being less apparent in naturally short-backed breeds of heavy type like Cochins and Brahmas.

Taken from *Fowls and How to Keep Them* by Rosslyn Mannering, published by Cassell and Company, 1924 all the other necessary factors are put in place as you will know by then exactly what you are looking for and even how much you intend spending.

In addition to organising shows (and sometimes selling their members' birds in a 'selling pen' or 'selling class' – of which more in Chapter 8) poultry clubs often organise sales and auctions. Remember that, although you will undoubtedly see plenty of good quality birds, it must be realised that a breeder is unlikely to sell off their very best stock, as they will need such birds in order to maintain their own particular strain and bloodlines. Sales held in the autumn, may, however, offer the option of buying older birds which the breeder does not want to keep over the winter months – while they might be past their prime as far as he or she is concerned, such stock is, nevertheless, still capable of producing show-quality offspring in the next rearing season. Also, purchasing adult birds at this time of year obviously has its advantages as they can be placed immediately in the chicken shed and run, rather

Ducks and geese rather than chickens and bantams are up for sale here – but no matter what type of poultry is on offer, there will always be a crowd of interested bidders

than requiring any intermediate housing or extra care and attention. If you are simply looking for birds that will lay plenty of eggs, then pullets hatched in the early spring of the same year have to be a better option than older hens, as you will obviously gain more eggs throughout their lifetime.

Rare breed sales are generally an excellent and safe way of buying stock as they are normally organised under the auspices of an accredited rare breed society that will only permit bona-fide sellers to participate. Alternatively, should you be merely searching for a pen of healthy layers – the actual 'pedigree' of which does not matter to you – your local weekly or monthly cattle mart might be the place to go. A somewhat less reliable method of sourcing your stock is via the various chicken-related blogs on the Internet – and here it must be made clear that this is not intended to include established breeders with their own websites, but more the average 'punter' with the odd pen of birds to dispose of. Although you may be lucky, there is always the danger of buying inferior stock which may also be misrepresented in the online advert.

Dry nostrils, alert eyes, a bright waxy comb and shiny feathers are all excellent indicators of a bird's overall health *(Photo courtesy of Elliot Hobson)*

How to recognise healthy stock

It is always a very useful asset (and not when just purchasing birds) to be able to identify the various signs typical of a healthy bird simply through its overall appearance. As well as the various points raised elsewhere, general indicators include the following.

- **Dry nostrils** – there should be no sign of mucus or discharge.
- **Bright, alert eyes** (check with the breed standard – if there is one – for correct eye colour variation). They should be clear, bright and well pigmented.
- **A bright, healthy-looking comb** – the comb and wattles must be bright and waxy in appearance; there are a few exceptions to this rule, such as when the breed is of an unusual comb type or when a hen has been brooding chicks, or is in the moult and not

On a subsequent visit to Audley Hall, I had barely arrived and had my bags taken to my room before Young James took me off to see his hens. His enthusiasm was still evident but now that they were being given access to the walled garden, an experiment which I suspect, most vexed the gardeners, his search for eggs no longer led him to the nest box, but to a secluded place under the raspberry cages where his birds had established a very cosy place indeed.

On later enquiry, I discovered that the birds had come from a Mr Cook in Kent and had been transported in wooden boxes via rail – a practice apparently most common amongst the brethren.

Taken from *Visiting Home* by J. E. Marriat-Ferguson, published privately in 1905

laying. Otherwise there should be no signs of lacklustre colouring or the inclusion of unusual-looking warty growths.

- **Shiny feathers in reasonably pristine condition** – if the birds have not yet moulted, you may notice broken feathers on the back and neck of females in a breeding pen, but this is probably caused by the mating cock bird rather than denoting ill-health. The underpart of the wings nearest the body is an area particularly favoured by lice and mites, so lift up the wings, gently brush back the fine feathers in the opposite way to their growth and look carefully for movement: although tiny, such parasites are visible to the naked eye.
- **Good weight and musculature when handled** – always try to handle any bird that you may be considering buying: even the breeds classified as 'light' must feel fleshy and well muscled. Check the breastbone – it should be reasonably well covered and certainly not sharp like a knife blade.
- **A clean vent** – inspect the vent (anal area) for signs of diarrhoea and for lice and mites, and also that there are no signs of sticky droppings or an unusual smell.
- **Clean legs and toes** – i.e. no lifted scales or strange-looking growths, which may indicate signs of the parasite which causes scaly leg. The toes should be straight, not curled or crooked.
- **General overall demeanour** – even in the close confines of the sales pen, birds should be seen periodically scratching in the shavings, feeding, preening or simply 'chattering' away to each other. Avoid buying birds from a pen where an occupant is moping around – it is generally a sign that all is not well.

Chickens coming home to roost!

A good strong cardboard box to which plenty of ventilation holes have been added is all that is required to bring your birds home. If, as is most likely, you are transporting them in the car, make sure that it is well ventilated and that the box is not placed in direct sunlight in the rear of a hatchback type of vehicle, especially if your journey is likely to involve a busy motorway and unexpected hold-ups might possibly occur. When stationary for even a short period of time, even the weakest of sunlight coming through the back window can be quite heat intensive, which will not do your newly acquired birds any good at all.

Once back home, gently lift your precious cargo from the box (now is most definitely not the time to discover that you have a phobia about touching chickens!) and place them into their new house with the pop-hole shut. In an ideal world, you would have brought them back at dusk, in which case, it is a simple matter to lift them onto their perch and leave them quiet until morning; however, the practicalities of life tend not to be quite like that.

The box you take your birds home in need not be elaborate, but it should be well ventilated

If it is still light, place food and water in the house and let them settle for a couple of hours before then opening the pop-hole and allowing them out into the run. You should resist encouraging members of the family or your neighbours to come round and look as it will have been a traumatic enough experience for your chickens without subjecting them to any additional stress. If all is as it should be, they will return to the house to roost as soon as dusk approaches – fowl are not, as has been proven on several occasions throughout this book, as 'bird-brained' as one might have supposed. However, should darkness fall and, with the aid of a torch, you discover that they have not gone to bed and are all crouching in a corner of the run (which is where they will be rather than in its centre, where an instinct for survival tells them they are more vulnerable to predators), it will be necessary to pick them up and gently place them on the perch. By the next evening, you should find that they are all safely ensconced on the perches and be able to drop down the pop-hole without any further problems.

GOOD PRACTICE

As accommodating as they are to the strange nuances of human behaviour, chickens, unlike us, do not appreciate a lie-in at weekends. Not for them a long, leisurely look at the magazine section of their chosen newspaper nor, at Christmas perhaps, pecking at a breakfast tray in bed before ambling downstairs at 11.00a.m. for a pre-prandial alcoholic drink. If you're going to keep chickens you must be prepared to create a daily, three hundred and sixty-five days a year, routine that is suited to both them and you. It is not, however, as onerous as it might seem because chickens demand little and, provided that the system decided upon becomes regular practice, they will adapt quite readily to your own particular lifestyle (and you'll be pleased to know that you do not have to forego those weekend lie-ins either – just so long as you're prepared to venture outdoors in your dressing gown and perform the morning duties before nipping back to bed with a mug of coffee and an armful of newspapers!).

The daily (and weekly) routine

When I first started keeping chickens as a child, one of them only had to leave a single dropping in the pristine shavings below their perch for me to be in with my little rake, shovel and bucket in order to remove it. Let them dare to enter the nest box without first of all wiping their feet and it would have necessitated a complete change of litter, so fearful was I of a dirty egg and the possibility of disease. Come what may, every Sunday was 'clean-out' day and my poor birds must have lived in dread of school holidays when I was forever shutting them into the house whilst I uprooted posts and wire netting to make a new outdoor run in another part of the orchard. Once the house was attached again and its occupants were allowed out to come blinking into the light, they must have been completely disorientated because their whole immediate surroundings had changed! When Newcastle disease appeared in our area during the early 1970s, every one of my birds suffered the indignity of being injected with a pellet under the skin, courtesy of a family friend who worked at a local poultry farm. On occasions, my wife dares to say that I am obsessive – looking at what I've just written, she might have a point and, in reality, the general routine of small-scale chicken-keeping is merely a matter of common sense.

On a daily basis, make sure that the feeders and drinkers are clean and contain *fresh* food and water. If the drinkers have, for one reason or another, to be kept in the house, make sure that any damp, soiled litter is removed and, while you are in there with your little bucket and shovel, remove any faeces from the droppings board (if your house is equipped with one) or directly under the perches. If the drinker is outside, move it to fresh ground so that the area surrounding it doesn't become muddy. Perhaps most important of all, adopt the traditional stockperson's stance and lean on the gate watching your birds – apart from the pleasure such time-wasting will give you, you will have the opportunity to see whether anything is amiss with any of your birds.

If your chickens are housed in a moveable house and run, depending on its size (and obviously the size of your garden), move it on either a daily or weekly basis. If it is of a stationary type, then, on a weekly basis, clear away any undue amount of droppings that have accumulated during the previous seven days and also remove to the compost heap any soiled cabbage leaves or similar greenstuffs that have not been eaten. Particularly in the summer, be sure not to neglect spraying or dusting any little nooks and crannies in the house (where the perches attach, the corners of nest boxes, etc.) where lice and mites might congregate. In any case, clean the nest boxes out once a week (more often if the weather has been bad and the nest is dirty because of the hens' muddy feet) and have a general check around to make sure the house has no loose nails, flapping roofing felt and that the run has no holes that have suddenly appeared either around the perimeter as a result of the birds' dusting activities or in the wire netting itself.

The daily 'practice' might not be pleasant – but it is necessary

OCCASIONAL CHORES AND GENERAL MAINTENANCE

As well as the daily and weekly cleaning routines, there are a few jobs that are specific to a particular season of the year. Large, traditional wooden houses (as opposed to the small moveable type which will perhaps need totally cleaning out once a week or once a month depending on how vigilant you've been with your daily routine) need a thorough cleaning out twice yearly: preferably on warm spring and autumn days. All of the litter must be removed – as should any house furnishings such as perches and nest boxes that can be unscrewed, or otherwise taken outside, for a thorough scrubbing. Your local agricultural supplier will be able to recommend a brand of safe, non-toxic disinfectant that contains an anti-parasitic medication and this should be used both inside the house and on any of the items you've removed. Let the house and fittings dry out completely before re-assembling things and allowing your birds access to their home. All houses will benefit from an annual treatment of preservative, which must, of course, be animal-friendly.

Some of the more modern designs of chicken house made from material other than wood require less general maintenance and repair *(Photo courtesy of Crofting Supplies, Caithness)*

On occasions, it may be necessary to clear paths and the immediate surroundings of chicken runs. While weedkiller is the obvious solution (and without wanting to be alarmist), great care should be taken because of the possible dangers of some types of weedkiller. The herbicide glyphosate doesn't degrade easily and can accumulate in soil and in the tissues of perennial plants, making them potentially toxic to birds or animals. Whilst a strimmer might be a good and very environmentally friendly way of ensuring that the grass and vegetation is kept clear of the run perimeter, be very vigilant when using one. This is not just for obvious safety reasons, but also because the cord or blade may cause unseen damage to the wire netting itself. You might not notice it, but your chickens might – and a fox most certainly will.

A HOLIDAY ROUTINE

When we first moved to France a decade ago, I never expected that I would be travelling as much as I currently am in connection with my writing and countryside consultancy work. As a result, I virtually filled the garden and orchard with chickens and bantams, some of which I've since had to move on because, even though I'm lucky enough to have a neighbour who will see to my stock, it's just not fair to ask him to be responsible for too much of it!

If you ever have to go away and doing so means that you will need to call upon the services of a family member or neighbour to look after the birds in your absence, it will pay to explain carefully as much as you can regarding the general routine. As well as explaining verbally, it does no harm at all to leave a typed list of jobs pinned somewhere in the feed shed. Make sure that the list contains the telephone number of a fellow fancier somewhere in the area – it is far easier for anyone concerned about a particular problem to phone and ask someone to come and give their opinion and help within an hour or two, than it is for the neighbour to ring you. The list should also contain the vet's telephone number and, if you are of a particularly efficient nature, you could even ring your vet before you go away in order to inform them that someone else will be looking after your birds for the next two weeks – this should make explanations easier in the unhappy event that your locum does, indeed, have to call the surgery when you are away.

You may even find it worthwhile colour-coding your feed bins, or at least identifying the contents of each by attaching a label or writing on the lid with paint or an indelible felt pen marker as an *aide memoire* to your neighbour. I know of one couple who use plastic bins in different colours to aid instant identification – blue for growers' pellets, red for chick crumbs, green for layers' and so on. They claim that, not only does it make it easier for their children when asked to look after stock, it is also useful to them when, on a Sunday morning, they venture out to feed after a heavy night on Saturday!

Colour-coded or not, galvanised feed bins are better than plastic ones when it comes to protecting their contents from vermin

6
HEALTHY AND HAPPY

A trio of very healthy-looking Buff Orpingtons

When I talk of 'healthy and happy', I am of course, speaking of your chickens, not yourself! Although, having said that, perhaps here is as good a place as any to point out that some humans (thankfully relatively few) can develop an allergy to feathers, or even a phobia of chickens – a condition known clinically as *alektorophobia*. Assuming that you have neither of these and chicken-keeping is, therefore, an option, I wouldn't worry overmuch about reading this particular chapter in too much detail because, with chickens bought from a reliable source and a regular routine established, it would be rare for the back garden fancier to have any health issues whatsoever. It is, dare I suggest, perhaps best that you simply treat the following pages as a reference, if and when anything appears not quite as it should be.

That bold statement is, however, based on the proviso that you know your stock well enough to be able to spot abnormal behaviour, no matter how mild. Fortunately, in a hobby of this nature, it is generally a simple enough matter to identify individual birds, and you will soon detect any signs that things are not quite right. These might include being away

> If the small poultry-keeper or beginner buys his stock from a good breeder, keeps all his houses and appliances clean, feeds on good and balanced foods and gives his birds plenty of fresh air in their houses he should have no troubles with disease.
>
> Should the birds appear sick...help should be obtained at once.... If only one bird dies or is sick and all the others are plump on the breast, do not think they will all die and dose them with drugs. Only the one bird may be sick.
>
> As long as your birds eat well, sleep well, look well and, if adult, lay well, they *are* well.
>
> Taken from *Starting Poultry Keeping: A Guide for Beginners* by the Advisory Staff of *Poultry World*, published by Iliffe Books Ltd, 1940

from the rest of the group, a droopy, hunched appearance, excessive thirst, lack of appetite, changes in the texture and shape of droppings, or loss of weight and condition.

Much has been written in the media regarding avian influenza ('bird flu'). It is a 'notifiable' disease, i.e. if it is found, the relevant authorities (in England and Wales, most likely to be Defra) must be informed. The only way it is ever likely to affect your birds is if they are unfortunate enough to contract it as a result of the disease being passed to them by wild birds, or via your laxness in not sticking to basic hygiene rules, when the disease has been reported to be within a certain distance of your property. If this happens, you should take precautions by not visiting local chicken-keepers, or, if you do, making sure that your footwear and clothing are thoroughly disinfected before seeing to your own flock. To make a great deal of any major poultry diseases would be unnecessarily alarmist and so I intend simply to outline some of the more likely minor health problems and scenarios that just *might* affect your birds at some stage during a lifetime of chicken-keeping.

COCCIDIOSIS

In a book of this nature, aimed mainly at the enthusiastic small-scale chicken-keeper, I would not ordinarily think it necessary to cover the subject of *coccidiosis*. However, as the labels of some food bags tell the reader that the mix includes 'coccidiostats', it might just be worth mentioning the potential problem – and why such medications are included. The disease normally affects young birds between perhaps three and ten weeks of age and typically shows itself as a loss of appetite, hunched appearance, feathers that look staring and stark, and probably most obviously, the droppings will be white and stick to the feathers or down immediately below the vent. The droppings may also be flecked with blood. The coccidiostats included in proprietary feeds will, however, most likely be enough to ensure that you never see any of these symptoms in a whole lifetime of hobby chicken-keeping.

BACK GARDEN PESTS

Rats are an ever-present problem wherever livestock of any kind is kept, and the environs of a non-moveable chicken shed, with its ready supply of food and suitable sheltering places, are ideal. Not only do rats eat vast quantities of food, they are also avid takers of eggs and newly hatched birds, as well as being known carriers of disease – some of which are harmful to humans. Their presence may at first go unnoticed but once a colony has established itself, their greasy-looking, smooth runs can very easily be seen. Provided that you keep the contents well away from your birds, poison baiting stations are the best option; they must, however, be continually checked and refilled, otherwise rats may build up resistance by taking insufficient quantities of poison. A Universities Federation for Animal Welfare (UFAW) working group has recently produced a leaflet giving advice about humane rodent control. In it they point out that rats and mice can 'suffer fear and pain, and…it is important that we should use control methods as humanely as possible.'

As to foxes, traditionally the best-known taker of chickens, a recent article in *Your Chickens* magazine commented, 'Even if you live in a very urban area, the fox is possibly the most dangerous threat to your chickens. Foxes are opportunists and…there has been a large increase in their numbers. A new breed of urban fox has no fear of humans; they live in towns and breed very comfortably, with people feeding them out of misguided kindness.' Your local council should be able to put you in contact with either a publicly funded or private pest control officer who will most likely use baited cage traps.

Wild birds will also take a surprising amount of chicken feed and there is a school of thought which suggests that they might be responsible for the transmission of some diseases likely to affect poultry.

Rats and mice will be encouraged by left-over food

Certainly more pleasant to look at than rats, the fact remains that songbirds will also steal a surprising amount of chicken feed. This can be prevented by using small mesh wherever practicable

Whether that is right or wrong (and, no matter, it is always a pleasure to see songbirds in the garden) it will pay, where possible, to have small enough mesh that prevents them from taking your precious pellets and grain.

Avian predators such as crows and magpies love eggs and chicks and, even in a suburban environment, will enter seemingly safe and protected nest boxes in search of them. Larsen traps are the best way of keeping them in check.

Last, but certainly not least, never forget that domestic cats and dogs are surprisingly efficient predators – sorting out that particular back garden problem is, however, likely to prove far more difficult!

STRESS FROM YOUR PETS

Research at Bristol University has shown that chickens suffer from stress. It can occur as a result of all manner of factors, but the most common one must surely be because of fear of your household pets. Cats might be happy to use the top of a coop and run as a sunbathing hammock, but their presence will undoubtedly alarm the occupants – especially if there is a broody minding her newly hatched chicks. Likewise, a young dog gambolling enthusiastically towards the chicken run out of curiosity will not do much good for your chickens' heart rate. Of course, given time, dogs, cats and chickens will all get used to one another and I have had birds that frequently came in through the open kitchen door to steal the dog's food – and what's more he let them!

Introduce dogs and chickens to each other carefully. If your chickens are new to your dog (or your dog new to your chickens!), take him quietly to look at them and chastise him if he shows any more than a curious interest. Believe it or not, cats can also be taught to show little interest in chickens and I know someone here in France who has her cats trained just as well as any Crufts obedience champion.

It should go without saying that you should never attempt to take a dog into the poultry section at an agricultural show but, despite the organisers' frequently placed notices requesting them not to, many people do.

FROSTBITE

Not only is frostbite a potential hazard to climbers of Everest, it can (rarely, it must be admitted) be a problem to chickens and, in the coldest, most exposed parts of the country, there is the possibility that it might affect the combs of your chickens. It is more often seen in large combed birds and shows itself as a dark discoloration. In fact, it is not the coldness of the night when the bird roosts that causes the problem, it's more to do with being in an overly warm house and then being exposed to the frosty elements when the bird goes outside in the morning. To prevent this, many knowledgeable chicken-keepers suggest that petroleum jelly should be applied to the comb of large-combed breeds in the severest of weather.

DUST-BATHS

One way of keeping your birds happy and healthy is to include a dust-bath in their run. They will quite often make one for themselves anyway – especially if the hen house is raised off the ground and the dry soil underneath is easily accessible, but in any case, it is an easy enough matter to provide them with one under a simple shelter. Use dry earth, sand, or riddled wood ash together with an occasional dusting of poultry insect powder and keep it all in place by surrounding the area with four boards forming a square. It can be as big as you like, but something around 1m (3ft) square which is filled with about 15cm (6in) of mixture will easy keep half a dozen birds happy and contented. And the reason

▼ Any dry sunny area will be soon commandeered by your chickens in order to avail themselves of a dust-bath

your chickens like a dust-bath? Well, it is primarily because of their natural instinct to rid themselves of lice and other parasites that might like to live in their feathers, but in addition, it also helps keep their plumage in good condition.

FLEAS AND MITES

At certain times of the year (mainly the summer and early autumn months), your birds can be subject to infestations of lice, mites, worms and fleas. Some are more serious than others, but all need attention and it will pay to get into a routine of checking birds periodically for any unwelcome guests. Inevitably, dark corners and the backs of nesting boxes will encourage mites and fleas to take up residence and so these must be routinely soaked with a dilution of 'Poultry Shield' disinfectant which is, according to David Bland, a leading expert in all chicken-related matters, 'the only solution which can penetrate the waxy egg coatings'. Alternatively, keep the house well dusted on a weekly basis with 'Diatom' which, being a natural product, has no ill effect on the birds. Remember that the types of mite which live permanently on the birds can easily transfer from chicken to chicken, or from a broody to her chicks.

Red mite

The red mite is a blood-feeding ectoparasite and, despite the name, is actually grey in colour, only appearing red after feeding on the blood of the host. They cause skin irritation and stress to birds and, in severe cases, will reduce egg-laying and egg numbers. Red mites live and breed in crevices found in the house and only use the chicken in order to feed – having said that, they can live for many months without feeding and have been known to lie dormant for several years. For this reason, always ensure that any second-hand housing you buy is thoroughly cleaned and disinfected before being used.

Just a few of the various mite preventative and curative preparations available without veterinary prescription

Northern mite

Northern mites are similar in size to red mites, and, like them, are normally grey to black in colour but will show as red when they have just fed. Unlike red mites, however, they will actually live on chickens and will probably be most obviously noticed around the vent area – although close inspection might also find them around the head (where they can, in severe cases, cause scabs to form on the comb, face and wattles) and under the wings at the point closest to the bird's body. Sadly, this particular mite is not as easy to eradicate as red mite and may well require several treatments with a spray or powder before the problem is finally solved.

Scaly leg mite

Scaly leg mite will burrow under the scales of the leg and breed in the cavities thus created. Petroleum jelly is known not only to block off the supply of oxygen to this particular mite, but also helps in loosening the old scales and strengthening the new as they grow. Some chicken-keepers suggest treating scaly leg with surgical spirit or similar alcohol-based liquid, but there are manufactured treatments on the market. Feather-legged types are known to be susceptible to this particular problem, which is obviously more difficult to spot amongst the feathers, and owners of such breeds need to be on constant look-out for its presence.

Lice

When large numbers of lice are active on a bird, the most obvious signs are greyish-white eggs around the base of the vent feathers. In really bad cases, they will also be seen around the 'armpit' area under the wings. The lice themselves are about the size of a pinhead and light brown in colour. Infestations are usually worse in the spring and summer months, causing irritation to adults and sometimes, severe anaemia in young birds. Fortunately, they are one of the easiest parasites to be rid of and, once discovered, can be eradicated by a proprietary louse powder purchased from your veterinary surgeon or, perhaps more easily, the local agricultural suppliers.

No matter what the cause, should any injuries occur they are, in the main, curable if noticed soon enough

FEATHER-PECKING

Several poultry equipment and feed suppliers sell pecking blocks, which, they claim, will help prevent feather-pecking because the food source thus provided in the blocks will keep chickens busy and entertained. So they will, but, with only a few birds in the back garden, you will most likely never see any signs of feather-pecking in your chickens because it is usually in intensive housing situations that it manifests itself. Gamekeepers and professional poultry producers, for example, occasionally experience the problem amongst their flocks and may have to resort to the fitting of bits or the trimming of the bird's top mandible purely because of the very large numbers they are rearing.

Symptoms often depend upon the age of the birds; in chicks it might be noticed as vent-pecking or on the tips of the newly emerging wing feathers, whilst in adults, the vent is often the place flock-mates choose to peck continually at the feathers of their unfortunate brethren. It is, however, important to distinguish between feather-pecking and the loss of feathers at the back of the head of some breeding hens which is quite commonly caused by the amorous attentions of the cock bird as he tries to mate with them. Should he eventually draw blood, though, it is possible that other flock members will see this and peck at it out of curiosity. Once they have the taste, they will continue to peck and chase the sufferer until she is trapped in a corner where they may continue until a great deal of damage is done – or they may actually kill the unfortunate bird. In the unlikely event of feather-pecking, and provided no infection is apparent in the wound, the application of a good quality anti-peck spray will prevent further problems and help heal the damaged area.

WORMS AND WORMING

Indications of the presence of worms vary but, generally, symptoms include a loss of weight – even though the amount of food being consumed is maintained – with diarrhoea, pink (as opposed to red) comb, and unusual lethargy. A regular worming programme is to be recommended but just how often this should be carried out depends upon the particular wormer being used. Although some veterinary products need to be licensed by the Veterinary Medicines Directorate (VMD), Flubenvet wormer is exempt from the ruling and can be used legally by back garden chicken hobbyists. Herbal wormers are also available without prescription, one of which is made by Verm-X who claim that it is 'available from over two thousand pet shops and country retailers'.

'Flubenvet' is a readily available wormer

Roundworms and tapeworms

Roundworms are round and smooth, whereas the tapeworm is segmented. The female roundworm lays her eggs in the intestines of chickens, which are eventually excreted before being picked up by other birds, hatching in the intestine of the new host and beginning the whole cycle again. Tapeworm eggs may also pass out of the bird via its faeces, or be retained in one of the segmented sections of the worm. These sections break off periodically and are also excreted, at which point they are sometimes eaten by small invertebrates who become the temporary host. If a chicken then eats that host, the cycle continues, but unlike the roundworm, if only the eggs or segmented sections are ingested by the bird, they will not develop into more tapeworms.

ISOLATING BIRDS

There is a lot to be said for having somewhere to keep your birds isolated on occasions – a kind of quarantine area if you like – where any new additions bought in, or birds brought back from shows, can be housed for a short period until it is obvious that they are not suffering from ill-health and have not picked up any minor ailments. If they are your own birds going back into a pen with their old friends, they need only be isolated for forty-eight hours or so, but if it is new stock, perhaps an

A series of small, outdoor, well-ventilated pens can be used as both 'quarantine' and show training cages

unrelated cockerel brought in to add new blood to your strain, then it should be kept isolated until one can be sure that it is not showing any untoward signs. Some show competitors also use the time that their birds are in 'quarantine' to give them a short course of multivitamins in their water, feeling that this will help them overcome any stress or exhaustion brought on by the show environment.

AUTUMN MOULTING

Sometimes a partial moult occurs in the early part of the year and is often limited to the neck, especially in the case of point-of-lay pullets that have been given a layers' ration too early in life, but generally, the feathers of most breeds of chicken are shed annually in the late summer or early autumn. Some find it protracted and difficult whilst others pass through the moult far more easily and quickly. It is certainly better for the bird if the moult happens quickly, and in order for this to happen, it helps if they are in good condition. Birds can look to be in a bit of a sorry state at this time – as the early twentieth-century poultry-keeper F. E. Wilson pointed out:

> 'MANY POULTRY-KEEPERS regard the process of moulting as a nuisance and neglect the fowls more or less while it is going on. This is a great mistake; birds should be carefully fed and otherwise attended to at this time, because moulting makes a considerable demand on the system, and the yearly casting of their feathers is a critical time in the life of all birds...there is some sympathy to be felt for the owner, for probably at no other time does the poultry-yard present such a melancholy appearance. The hens mope about in corners and look generally ragged and untidy; while the much desired eggs are few, and in many instances altogether wanting.

IMPACTED GIZZARD

If you drop and leave a piece of garden string or baler twine in your chicken run, it will not be many minutes before it is pecked, frayed and quite likely swallowed by the birds. It is therefore essential to pick it up immediately. Provided that chickens have been given access to sufficient hard flint grit (which resides in the gizzard and acts as a grinding material in much the same way as a mill-stone turns wheat into flour), nine times out of ten all will be well if they attempt to pick up naturally occurring things such as long grass. Sometimes, though, problems do manifest themselves – but the typical symptoms are impossible to describe and the condition is difficult to diagnose. It therefore makes sense to ensure that (with the obvious exception of floor litter, about which one can do nothing) chickens are prevented from having access to any potential gizzard-impacting material.

A blue Araucana looking as if it's about to eat something it shouldn't. It pays to keep the run free of discarded string or anything that might be potentially harmful

Antibiotics and human safety

There are a few antibiotics where one is advised not to eat the birds' eggs for seven days after completing a course, and others for which there is no withdrawal period. The main reason for having withdrawal periods at all is as a safety precaution, rather than there being any serious danger – unless, of course, antibiotics were being administered continuously and were, in turn, ingested by humans. Not that many years ago, large-scale turkey producers had small amounts of antibiotics included in the feed as standard practice and that may have led to a risk. If an owner wishes to be on the safe side and takes antibiotics themselves on a regular basis, then a seven-day withdrawal period may be sensible.

LAMENESS

Lameness can have several causes and these have, in older books, been subdivided into categories such as 'bumble-foot'. There can be many causes of lameness, and a hot swelling on the foot pad and between the toes could be the result of infection caused by a wood splinter or thorn which, if it can be seen and removed and the wound dressed with an antibiotic spray, should cause no further problems. Swollen pads may also be the result of jumping down repeatedly from too high a perch onto hard ground, and a limp may simply be the result of a bird catching its foot or straining its leg as it scratches around in search of natural food. Chickens can also suffer from arthritis, and there are two kinds – acute and chronic – as there are with humans.

A poultry pedicure

If your chickens are kept out of doors and with plenty of scratching material available, it should be unnecessary to trim their toenails. Occasionally, though, one can become damaged and it is a simple matter to rectify the situation with a sharp pair of nail clippers designed for humans or dogs. By holding the foot up to the light it should be possible to see where the blood vessel or 'quick' ends and the actual nail begins. It is important not to cut into the 'quick' otherwise it will bleed quite badly and require cauterising.

ENDING A CHICKEN'S LIFE

Just how long a chicken can live a natural life varies enormously; I would say eight years, but I have personally had several chickens and bantams which have lived until the ripe old age of thirteen or more. It is a fact that, where you have livestock, you will, sadly, eventually, have deadstock; your chickens might be happy and healthy throughout the whole of their lives but no matter how much care and attention you give them, they cannot live forever. There will come a time when a bird begins to lose condition and become lethargic, it may stop eating (but most will continue to drink) and suffer from the 'pecking-order' in that it starts to be bullied by the others and, as a consequence, takes itself off to mope in a corner.

As with a cat or dog, it will perhaps be your natural inclination to take the chicken to a veterinary surgeon. In my opinion, the trip is not necessary – even supposing that you can find a vet willing to undertake the task on your behalf – it is far better to kill the bird quickly and painlessly yourself by dislocating its neck. It is generally a simple matter: hold the bird firmly by its legs in your right hand and cup the top of the back of its neck between the 'C' shape formed by the thumb and index finger of your left. If done correctly, a quick pull downwards and simultaneous flick backwards will kill the bird within a fraction of a second. Although easy to describe, it may not be so easy to visualise and so it is always better to ask a more experienced poultry-keeper for advice. Hopefully, when the time comes for one of your birds to be killed, a member of your local smallholding group or chicken club will be prepared to do the deed on your behalf and, although it may prove difficult to watch, you should do so in order to gain vital experience.

If this is correctly done, death will be instantaneous; you will, however, notice that there is an inordinate amount of wing-flapping and body-twitching. This is caused by simple nerve reflexes and, distressing though it may be to witness, you can rest assured that the soul and character of the bird you have known and loved has departed several seconds before.

7 BREEDING, INCUBATION AND REARING

A great deal of the fun to be had from keeping a few chickens in the back garden is the prospect of one day hatching a nest box full of fluffy, healthy-looking chicks. The process is relatively simple and straightforward; however, should you choose to breed to standard and eventually exhibit the offspring of your birds, some breeds can, because of complicated genetics, be a little more difficult than others. The details of specific breed colorations and correct feather markings are beyond the scope of a general book such as this, but there is plenty of help to be found on the Internet (look at a particular breed's club pages), via the Poultry Club of Great Britain and in various specialist magazines. Of course, the best place to seek assistance is from the person who supplied you with your initial stock birds – there can be no substitute for personal experience and they will, in the past, have made mistakes and learnt a great deal as a result, all of which will undoubtedly prove beneficial to you. Never be frightened of asking what you might fear to be 'stupid' questions; if you don't know the answer and need to, then it's not a stupid question!

How can anyone resist the allure of newly hatched chicks?

SETTING UP A BREEDING PEN

In all likelihood (and unlike serious breeders who will put together a pen in late December/early January in order to breed their birds in the spring), you will keep your hens and a cock bird together all the time and so there is no need to worry about how long a male has to run with his females before the eggs are fertile. Generally, though, it is necessary to ensure that the cock bird has been with his hens for a least a month before fertile eggs are needed for hatching, as, although a cock bird will ejaculate enough semen into the hen's oviduct to fertilise a dozen eggs or more at one mating, sufficient time has to be allowed to ensure that mating has actually taken place and that the resultant sperm has reached the oviduct. It might also be worth saying that when hens have been running with a cock bird who is unsuitable for breeding and it is decided to replace him with another (maybe of a better standard), eggs should not be incubated until the new male has been with the females for the same length of time because sperm can remain viable in the hen's oviduct for around three weeks and there is, therefore, a very real danger that any eggs set before this period may have been fertilised by the original cock and not the replacement. However, as has just been mentioned, the majority of back garden chicken fanciers keep their flocks of birds penned as a breeding group all year round and so the eggs should always be fertile – although extremely cold weather in the early part of the year may affect the fertility of both male and female birds.

The precise timing of when a cockerel becomes a cock and a pullet a hen, is mentioned in the *Glossary*, but most experienced poultry breeders reckon that, as a general rule, a cockerel is such until he has finished his first season of mating (which might be as old as 18–24 months), while a pullet becomes a 'hen' after her first season of lay and not before. They also say that the best chicks, as regards hatchability and subsequent quality, come from a first season male mated to a second season female. True though this observation may be, it is not necessarily an option open to the hobbyist who will, of necessity, simply have to breed from

Pullets may be used for breeding, if well-grown and not less than eight months old. The results from yearlings or hens are, however, much more reliable. Nevertheless, all prospective breeding females should be individually handled. Birds, in order to be retained, should be of good size for their breed and should be reasonably near in colour or markings to the breed they represent. They should be fit, as shown by the state of the eye and face, and the general body condition should not be too fat or really thin – a 'hard' condition is best.

Care should be taken not to include any birds which are rather 'beaky' looking, since this is often the first outward sign of impending uselessness.

Taken from *The Smallholder's Pocket Guide* – edited by S. A. Maycock, published by C. Arthur Pearson Ltd, 1950 what they've got – and, provided the birds are healthy and of a good 'type', there are no problems with this.

The cock-to-hen ratio

One should, of course, be wary of breeding from birds that are too closely related to each other because, although 'line-breeding' and 'in-breeding' are practised by some experienced breeders in order to accentuate a particular positive point, or out of necessity because there are few birds of a certain breed about, they are devices best avoided by the beginner. So, when buying your birds, try to ensure that the breeder sells you unrelated stock or perhaps even go to the trouble of buying your hens from one supplier and the cock bird from another.

Can there be anything more perfect than this stunning group of Partridge Wyandottes? Their conformation and adherence to breed standard make them ideal breeding material

How many females should run with one male is generally a 'done deal' because you will most likely have bought yourself a trio (one cock and two hens), a quartet (one cock and three hens) or perhaps half a dozen laying birds to which you decide to add a cock later. Lighter Mediterranean breeds such as Anconas, Leghorns or Minorcas will cope with seven or eight hens to one male, but heavier breeds such as Jersey Giants or Brahmas are best with two, three or four hens per male. Traditionally, outdoor commercial flocks of Rhode Island Reds or Light Sussex would have been flock-mated – that is, several cocks were run with quite a large flock of hens and there could have been as many as eight to ten hens for each cock bird.

EGG COLLECTION AND SELECTION

Contrary to what some people believe, it is not necessary to keep eggs warm prior to incubation. Collect them as often as practicable and store them in a cool place until you have sufficient for the broody hen or incubator. Properly stored, eggs will stay viable for quite some time; it is generally accepted that they should be no more than a week old if being hatched in an incubator, but the time can be extended to 10–14 days if they are hatched under a broody. However, the fresher the egg, the better chance it has of producing a strong healthy chick that hatches on time (having said that, do not set any eggs that are *less* than twenty-four hours old). Mark the date an egg was laid on the shell with a soft-leaded pencil so that you can always be sure of an individual egg's age.

Freshly collected Orpington eggs

An obviously misshapen and flawed egg which shouldn't be used for hatching...

...like this particularly poor specimen!

Breeding for egg colour

Purely out of interest, if you wish to breed so as to create the deepest possible brown egg, the basic way is to breed from birds that lay the darkest eggs over the longest period. Such a project may take a great deal of time and could result in keeping more stock than you would normally want in order to ensure that you are only breeding from those that consistently lay the darkest of eggs. In addition, although it is known that it is somehow related to genetics, the breed alone does not guarantee a consistent supply of continually dark brown eggs. A male bird could hatch from a dark brown egg, but may not have the genes to father pullets that produce dark brown eggs – it seems to be accepted by most that the father is the determining factor for the colour of eggs in the next generation. Eggshell colour is polygenic, meaning that there are multiple genes that affect the end result. The particular pigment that is responsible for creating a brown eggshell in the oviduct is a haemoglobin porphyrin, in effect a blood product. Eggs with a blue shell, such as those from the Araucana, are created by a zinc chelate of biliverdin, which originates as a result of a compound synthesised in the liver.

Apparently, if you cross an Araucana with a brown egg layer it's possible to end up with offspring that lay a pinkish-coloured egg. It could be an interesting project to mate different but compatible breeds of birds in order to achieve a hybrid which consistently produces the perfect coloured show egg – but that might be taking things a tad too far (*see also* the comments about cross-breeding in Chapter 10).

When pullets begin their laying life, it is not unusual to come across irregular-shaped eggs, but only the most perfect eggs should be used for hatching purposes. Any that are misshapen, or whose shells are chalky or lumpy in appearance or texture, must be discarded and used in the kitchen rather than set under a broody or in the incubator. Having said that, I must mention that, many years ago, I had a bantam hen who always laid over-large eggs which were always chalky in appearance and once, desperate to make up a clutch to go under a broody, I included a couple of hers and they hatched. I then used her with a prize-winning cock bird and almost all of their subsequent offspring went on to be show winners – as an old editor of mine once said, 'never state anything categorically when talking of nature because it has a habit of proving you wrong!'

The best method of cleaning eggs for hatching is to wash eggs in a proprietary solution. There are several on the market and all, the manufacturers claim, will kill bacteria on the shell surface as well as a fair proportion of those that may have lodged on the outer membrane having gained access through the egg's pores.

SETTING EGGS

'Setting' is simply a term many breeders use to describe the action of placing fertile eggs either under a broody hen or into the incubator. As to whether hatching should be carried out with a broody, or with the aid of an incubator, there are plus and minus points for both – personally, though, I think that the broody has to be the best option for the enthusiastic back garden chicken-keeper who only ever needs or wants to hatch a single clutch of eggs each year.

A broody will save you the bother of worrying about humidity, airflow and regular turning of the eggs – all of which are serious considerations if you decide to use an incubator. She is also unlikely to succumb to an electrical power failure but, nonetheless, no broody, no matter how docile and trustworthy she has previously proved to be, can

Eggs ready to go!

be guaranteed to sit on her eggs for the full duration of the incubation – which can be very disheartening if she gives up with only a few days left before the chicks are due to hatch. There is also the difficulty of finding a broody; this is likely to be less of a problem if your chosen breed is one known for its propensity to go broody at the drop of a hat (usually these are the heavier types), but could prove difficult if your breed is known for its laying abilities rather than for its inclination towards motherhood. In such cases, some breeders keep a pen of birds purely to use as foster broodies, but this is not likely to be a practical option in the back garden where space is already limited.

Modern incubators, even small ones, are quite sophisticated and, provided that the manufacturer's instructions are followed closely, relatively easy to use. Their most obvious advantage over broody hens is that, once they are set up in a suitable room or outbuilding, they can accommodate more eggs per setting. They also do not need feeding on a daily basis – although they will need checking and, if it is a model that doesn't include an automatic system, the eggs will require turning twice daily. One enterprising company even hires out incubators – although if your hobby is likely to prove long-lasting rather than just be a five-minute wonder, it will undoubtedly prove cheaper to buy your own. The company's main intention is, I believe, to offer a form of biology lesson to schoolchildren, and for this reason supplies incubators that are fully automatic and transparent. As their website says, 'Our incubators are especially chosen to give the maximum viewing space for the children. The wonderment and excitement of watching a chick hatch is magical for everyone, but children are often transfixed and a good view is [therefore] essential.'

How should eggs be hatched?

As so much is heard about incubators in these days, one might well suppose that the services of a broody hen have been almost done away with. As a matter of fact, the work of hatching…is about equally divided between hens and incubators. Highly useful, and indeed indispensable, as incubators are when poultry rearing is carried on to any material extent, the sitting hen cannot be beaten where the rearing of prize and other valuable stock has to be done in winter and early spring. The chickens are stronger and healthier, for the natural heat and comfort derived from the hen is better for them in every way than artificial heat and nursing.

Taken from *Poultry Keeping and How To Make It Pay* by F. E. Wilson, published by C. Arthur Pearson Ltd, 1903

Caring for a broody hen

As a youngster, if I didn't have a broody hen of my own when I needed one, I used to find someone who had, and pester them until I was allowed to borrow it! At one time, gamekeepers relied on foster hens to hatch their pheasant and partridge chicks and would make early season arrangements with any local poultry-keepers or farmers. They would undertake either to buy broodies from them at 10 shillings (50p) or hire them for the sitting and hatching period for 2s 6d (12½p). Sadly, the best one can hope for nowadays is to do as I did and borrow a broody for the duration.

A good broody should be obvious: it will be seen clapped tightly in the nest box and, as you approach, will fluff up its feathers as a defence mechanism. If you gently slide your hand under (palm downwards), she will try and peck the back of your hand and maybe even shuffle your fingers with her feet as if she were trying to turn eggs. A hen which is simply laying an egg may also do all these things (although it is more

A perfectly happy broody hen (note the cover round the front of the box which helps to keep the nest in place and the eggs from falling out)

likely that she will noisily vacate the nest with a very disgruntled expression on her face) so it is important that you observe the bird in the box over a period of two or three days.

Looking after the sitting broody

The hen will need a quiet secluded place in which to sit her eggs. If at all possible, a small coop and run set well away from the distractions of the rest of your little flock will provide a home for her while she sits, and is also the perfect place for her to bring up the chicks. Alternatively, if one has a penning shed (*see* Chapter 8) she could go in there – if not, it will have to be a broody box in a suitable corner of a vermin-proof garden shed, garage or outbuilding (it must not, however, be subject to fluctuations of temperature).

A nest can be made by first of all placing an upturned grass sod, cut to the correct dimensions, in the base of the broody coop or box and a saucer-like depression shaped in its centre (this will help retain necessary humidity). Next bank up the nest and sides of the box with suitable material – this would, at one time, invariably have been straw or hay, but there are now cleaner, more hygienic alternatives to be had – making sure that the corners are packed tightly and there is no way that eggs can roll out of the nest. For the same reason, the front of the box should have a quite substantial lip tacked across it. Once you are absolutely certain that your chosen hen is broody, place a couple of dummy eggs in the nest and then, in the evening, gently move her onto them and give her the following twenty-four hours to settle before very quietly and gently taking away the dummy eggs and replacing them with your precious (hopefully) fertile ones.

A good broody knows when her chicks are due to hatch and will sit in the nest despite the first few demanding her attention

Depending on the hen (some take their duties so seriously they will not leave the nest of their own volition), you can choose either to keep the box closed and remove her morning and early evening in order that she can feed, drink and empty herself, or place a supply of food and water in front of the box and leave her to her own devices. If you choose the former option, use great care when lifting her off the nest to ensure that she does not have any of the eggs caught in her feathers, otherwise they may fall and break as you bring her out. As incubation progresses, you will notice that she loses some of her breast feathers; this is quite normal and the area is known as the 'brood spot' – its purpose is to ensure that the bird's body heat is efficiently and correctly transferred to the eggs.

A couple of days before the chicks are due to hatch, the hen will hear them cheeping and be very reluctant to leave the nest for either food or water. Even if you lift her out, she will, in all probability, be anxious to return to the box rather than feed or drink. A little aired water mist-sprayed on the eggs and nest at this time will help to ensure that there is sufficient humidity and that the shell membrane is not too dry, which will make it difficult or even impossible for the chicks to 'pip' their way through the shell.

The majority of chicks should hatch together: you may be lucky and all will hatch, but it is quite likely that some will not – this could be because the eggs were infertile ('clear'), or the chicks had started to form and then, for one reason or another, died in the shell. If you gently shake the egg at this point, you might hear the faint cheep of a chick yet to hatch – if the egg makes a liquid sound, it is 'clear' and can be thrown away. When hatching in a coop and run, once all the chicks are dry, put them and the broody in a cardboard box and quickly clear away the nest before replacing it with a covering of clean wood shavings and allowing the hen and her chicks to return. If they are to be cared for elsewhere, take them there and let them settle in their new quarters.

Hatching with an incubator

As mentioned previously, it is very important to follow the manufacturer's advice to the letter when hatching with an incubator. For this reason, it is pointless going too deeply into the specific details and a few general points will suffice in this particular section.

Incubators for the 'big boys' – far more suited to commercial chicken rearing than the back garden; the basic working principles, however, remain the same

- If the incubator is not equipped with an automatic turning system the eggs will need to be turned twice daily by hand in order to prevent the developing embryo from sticking to the side of the shell.
- The incubator should be housed in a place that retains a constant room temperature because, if it is subject to the variations caused by fluctuating central heating or outside temperatures, the likelihood of a successful hatch will be lessened.
- A good air-flow throughout the whole of the incubation area is also important – stale air in the room will affect the results. Adequate ventilation is essential because the machines take in fresh air and give back stale.
- Interesting as incubator hatching undoubtedly is, try to avoid too much messing around and generally behaving like a mother hen! The thing most likely to reduce the hatchability of perfectly viable eggs is the undue interference of the incubator operator.
- The final sequence is this: a couple of days before hatching, the chicks will start to cheep. A day before the chicks actually hatch, they will make a first crack in the shell and it should be possible to see a tiny beak resolutely working to make the hole bigger. Eventually they will break the shell open along a very neat and tidy line. When first hatched, they are very weak and somewhat slimy looking; however, it will only be a few hours before they have completely dried and are ready to be transferred to the brooder.

BROODING CHICKS

A broody hen will be quite happy to squat with her chicks in the same position as she's been during incubation for perhaps as long as a day before she begins encouraging them to feed and explore. She will then squat down again periodically and all the chicks will disappear under her for a quick warm-up. There is no need to worry about the youngsters

 Chicks under a heat lamp

not feeding during these first few hours because they will be getting enough sustenance from the yolk sac they absorbed immediately prior to hatching. The hen will look after her brood very well from now on – all you need to do is supply a safe environment free from vermin and disturbance, some food and water and let her do the rest. It is, however, a very different matter if the eggs were hatched with an incubator and the chicks need artificial brooding.

Artificial brooders

It makes sense to buy a brooder from the same place as you buy your incubator. There are many variations on the market, but the purpose of all of them is exactly the same: namely, to keep the chicks warm! As with the incubator, it is important to carry out the maker's instructions to the letter. One of the simplest ways of providing heat is to use an infra-red heater suspended over an enclosed area (the heater can be equipped with either a dull-emitter or a special bulb – the latter is, I think, best as it gives off a warm cosy glow which seems to encourage chicks to stay in its light).

Turn the brooder on twenty-four hours before the chicks are due, so that the immediate area will have aired thoroughly and there will be some ambient warmth – remember that your chicks will have just come out of an incubator with a running temperature of around 39.5°C (103°F). Make subsequent preparations depending on the type of brooder you have: with the hanging infra-red variety, for instance, it should be suspended about 45cm (18in) above the floor. When the chicks are eventually introduced, allow about half an hour before you decide whether or not the heat source requires adjusting. If they are too cold, the chicks will be standing on tip-toe in the centre and cheeping very loudly; if too hot, they will settle in a large ring around the sides. If, however, they spread comfortably under the heat source leaving a small clear area in the very centre, they are, like Baby Bear's porridge, 'just right'!

FROM DAY-OLD TO SIX WEEKS

The makers of a specific brooding unit will, in the instructions that accompany their product, give advice on weaning your chicks from the heat. If you've chosen the simple, yet very effective, means of brooding with infra-red lamps, you should have been gradually raising the heat source a little each week so that, from an initial temperature of about 32°C (90°F), at the end of the third week the temperature is down to around 21°C (70°F). There are, of course, no such worries when hatching and rearing under a broody hen as she will do everything required as a result of thousands of years of instinct.

At three weeks, the youngsters should, depending on the breed, start

Feeding chicks

Don't expect your chicks to eat very much for the first 24–48 hours as they will still be living on the absorbed yolk sac. From then onwards, their appetite will start to increase so it is important that they always have feed available. Do not, however, feed so much that there is unnecessary wastage as, apart from the expense of it all, spilt food will very quickly become sour as it lies in the floor litter and becomes damp or wet due to inevitable spillage from the chick drinkers.

Your feed supplier will advise you on what type of food will best suit the age of your chicks. At first they should be given chick crumbs followed a few weeks later by growers' pellets, and, eventually, they will move on to adult rations. Chapter 5 gives more details of additional foodstuffs, the types of feeders and drinkers available for chicks, and also covers the subject of hard flint grit.

Six-week-old Silkie cross chicks with their 'mum'

growing proper feathers on their backs and around the vent area. As the wing feathers develop more quickly than the rest, the chicks are already gaining some protection against the elements, until by the age of about six weeks they are fully feathered and could, if required, be removed from the broody hen or artificial heater. Personally, I would advocate encouraging the hardening-off process for artificially reared chicks by turning off the heat source during the day from about three weeks of age but, of course, much depends on weather conditions at the time.

By six weeks, it should be possible to identify the sexes – the cockerels will have larger combs and have a different stance from the pullets, and, although they will not have an adult tail, its final shape will be well on the way to being defined. What you do after this point is entirely up to you – while you will be able to put the once-broody hen back with the flock, it is most certainly not a good idea to return her with her chicks as, despite possibly being related, the other adult birds will find their presence disturbing and may even kill them in an effort to re-establish the 'pecking order'. Once you have reared a clutch of chicks – be it through natural or artificial means – you have established a new generation, all of which will need separate accommodation and extra space. No wonder chicken-keeping is a 'growing' hobby!

8

A LITTLE BIT OF SHOWING OFF!

A view of the National Show, held annually towards the end of the year

The popularity of showing has to be seen to be believed; at the last Federation Show held in Staffordshire, it was reported that there was 'a record entry of exhibits', whilst a short while later at the Scottish National, 'they once again had an increased record entry of 3,574'. That's a lot of chickens and bantams by anyone's standards! Even if you never have any intention to show your own chickens, you will be missing a great treat if you don't go along at least once in order to watch others exhibiting theirs. Whether it is organised by a local poultry club and held in a village hall, or is a prestigious national event where, in order to be able to accommodate all the expected entries, the venue might be one of the UK's great exhibition centres, they are all worth attending – especially if you've yet to make a final choice on what breed of chickens you'd really like to keep and you're still looking for ideas. Not only will there be a good gathering of varieties all under one roof, there will, especially at some of the larger shows, be breed clubs and their representatives, all of whom will be happy to answer questions and offer advice.

SHOW 'POLITICS' AND ADVICE

Even at a small, local show, it would be rare to find exhibitors, stewards and judges who are unwilling or unable to offer assistance. Poultry showing of any description is, arguably, one of the friendliest pastimes and, unlike some other types of livestock exhibiting, is not generally beset by any bad feeling, ill-grace or the attitude that 'well, there's no point in going to that one…so-and-so is judging and they only like big/small/light/dark animals.' At some livestock shows (and, on more than one occasion have I heard it said), a qualified judge who happens to be exhibiting on that day has been heard to comment to the judge of the moment, 'if you give mine first prize today, I'll do the same for you the next time I'm judging.' It may, of course, simply be the degree of anonymity that prevents the likelihood of such a scenario: it is far more difficult to know who owns what bird because, unlike a dog show, for example, the owner is not likely to be anywhere near their fowl when judging actually takes place. The birds are penned; the exhibitors are generally ushered out and the judges are left to make their own decisions without hindrance or influence. Another point to make is that, in my opinion, there is less leeway for a judge to allow their fondness and liking for a particular size or colour to take precedence because poultry standards are, dare I venture to suggest, far more tightly regulated than those for many other types of livestock.

Advice from the past

To be successful, preparation is all-important: you cannot expect simply to open the door of your chicken house the night before an event, select a bird at random from the perch and expect to carry off 'Best in Show' the following day. The selection of a potential winner and show preparation is an art which takes time and patience to learn. It is, however, within the scope of most, but perhaps not all – as Rosslyn Mannering remarked in this particular paragraph of her book, *Fowls and How to Keep Them* (published in 1924 by Cassell and Company):

'Even if your bird fails to win you will be able to compare it with the winners and find out where yours fails, and – one hopes – excels, although the exhibitor who sees nothing but perfections in his own birds, like the consummate egotist who is oblivious of his failings, damns his chances of ultimate real success and triumph at the outset.'

Some twenty years previous, poultry writer F. E. Wilson also advised the newcomer to showing the ill wisdom of thinking that:

> BY PURCHASING at perhaps a stiff figure a winning bird you are going to carry all before you. The thing is not done like that. In the majority of cases it takes years of patient and hard won experience, keen observation, and the faculty of learning to profit by the mistakes either of yourself or others, together with that indefinable 'something' that marks the fancier who has succeeded or who will succeed – a sort of insight, oftener innate than acquired – to touch the pinnacle of success, and breed birds so good that no one, either prejudiced or unprejudiced, can gainsay their merits.

PENNING SHEDS AND SHOW CAGES

There are plenty of extremely successful exhibitors who manage quite well without having the benefit of a penning shed. A penning shed is, at its simplest, a building (a well-constructed garden shed is ideal) which has good ventilation and light and is equipped with four or five show cages, the sizes of which will depend on whether you are showing large fowl or bantams. With solid floors, they can either be tiered, or run lengthways along the room. Although it is possible to buy show cages for this purpose – suppliers for which can be found either on the Internet or in *Your Chickens* or *Fancy Fowl* magazines – home-made boxes fitted with wire fronts will do equally well. It doesn't have to be a beautifully crafted piece of engineering and/or carpentry as the main point is to accustom your bird to feeling happy and at home. It is of course important that your pens have food and water vessels attached to the bars rather than loose on the floor because it is likely that at some stage when chickens are in your practice pen they will have been prepared for exhibition, and the last thing you need is for their plumage to become soiled.

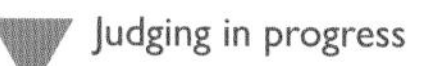
Judging in progress

Whilst accepting that such a set-up is not essential, there can be no doubt that as far as training for future shows is concerned, young stock benefit greatly from being in these pens and will stand and behave more confidently as a result. The judges handle each bird in order to assess its conformation, colouring and much else besides, but if there is a situation whereby two particular birds have amassed the identical number of points, it will undoubtedly be the one that has stance and bearing which will be placed first.

'TRAINING' SHOW STOCK

Over one hundred years ago, Edward Delaney, an Edwardian exhibitor, wrote:

> IN SHOWING POULTRY, much depends upon whether the birds have been properly trained or not. They must be thoroughly tame and have perfect confidence in you. If an untrained bird is placed in an exhibition pen, he is frightened by being handled by the judges or attendants; he loses all 'form' and style, and as often as not, huddles himself down in a corner and cannot be induced to show himself. This is fatal to his success, as may be imagined. The trained bird, on the other hand, is fond of being handled, and appears to be proud of showing himself off to advantage. The best plan is to begin with the chickens, getting them perfectly tame, accustomed to being handled or touched with a stick. If their natural confidence has never been abused, there will be no trouble whatever with them.

Very little has changed in the intervening years. In a very short time, your birds will become used to you going in and out of the penning shed and brushing past in close proximity, all experiences which will help them to be confident on the day. Judges very often carry a small telescopic stick (much like a car aerial) which they sometimes poke gently through the bars in order to get a show bird to stand properly. You might like to keep a short, thin garden cane by the side of your show cage and occasionally do the same – it all helps, and such simple things can make the difference between success and failure. Periodically take the bird out of the pen, handle it, gently spread its wing feathers and even blow into the downy parts of any plumage – all things that the judge will do. You could even accustom it to standing quietly on a box or table top in preparation for when you win first prize and a photograph might be needed for inclusion in the poultry press!

It will pay to train your birds to stand well in preparation for when you win first prize and a photograph will be needed for inclusion in the poultry press!

Training tit-bits

Back in Chapter 5, I was quite adamant that, in most cases, giving little 'treats' to your chickens was quite unnecessary – except, that is, when preparing a bird for the show bench. Oily foods, such as linseed or sunflower seed, are sometimes used to help the gloss of the plumage and a little maize included in the food will help in giving both a healthy colour to yellow-legged birds and a golden yolk to any eggs you may be thinking of exhibiting. You cannot, of course, simply give these on one day and expect immediate results the next, and it is necessary to feed any of these additional foods for perhaps a fortnight beforehand.

Where tit-bits really come in handy, though, are as a means of hand-taming and quietening your chickens and bantams – few can resist pecking at a little wild bird seed or dried meal-worms! Some of the traditional showmen would use hand-feeding as a means of getting any breeds that should have a tall elongated stance (Modern Game, for example) to reach to the top of the pen by first getting them to associate a human hand with food – eventually a little treat dropped into the training show cage from above would have their stock reaching up in anticipation, from which point it would not be long before a bird began to associate people around the pen with a tasty snack and would automatically stand to their best advantage.

You can imagine from the stance of this particular example just how the giving of a few tit-bits from above will have birds reaching heavenwards in anticipation

GENERAL SHOW PREPARATION

There are many little tips and wrinkles in use by the show fraternity that will enhance a potential prize-winner and give it the best possible chance when it appears before a judge (not really that much different from a human donning their best clothes before going to court for a minor motoring offence!). While chickens and bantams might not yet be subject to drugs tests, as would be a racehorse or athlete, there are some enhancements in the exhibition world that are legal and some that are not. The expression for the latter is 'faking' and you are likely to be disqualified from a show if, for example, you attempt to disguise any imperfections in a bird's plumage. Little tricks that are permitted include the use of Vaseline on the feet, comb and wattles to enhance the colour, smoothing trimmed beaks and nails with an emery board and, most commonly of all, washing a soft-feathered bird a couple of days before it is due to appear on the bench. (See, I told you it was like a human court appearance!)

Way back in 1903, F. E. Wilson remarked in his book *Poultry Keeping and How to Make It Pay* that, when showing birds:

> THE MAIN THING is to keep the birds clean: there must be no soiled bedding or litter in their pens, or their plumage will suffer. White fowls have sometimes to be washed, and the best way to accomplish this is to use a large tub or pan filled with warm water. Use a sponge for the plumage, and a small brush for the shanks and feet. Soap flakes are, of course, necessary, and every particle must be removed from the feathers, or the bird will look worse instead of better after the operation. The rinsing – which should be very thoroughly done – should take place in very slightly tepid water, in which the merest suspicion of laundry blue has been put. The fowl must then be put on clean litter and allowed to dry in a warm room. The legs are often rubbed with an old silk handkerchief…and for birds other than white fowls this oily handkerchief is rubbed over the plumage. Some birds require very little getting up if they are, as they should be, in the pink of good health.

Bath-time!

Traditionally, as F. E. Wilson commented, one used to wash chickens for the show in pure soapflakes and then, if they were of a white plumage,

Bathing a soft-feathered bird is generally considered to be normal practice

give them a rinse in a bowl into which 'dolly-blue' had be added. Nowadays, it is common practice simply to use lukewarm water and a children's hair shampoo (or an anti-mite dog shampoo). As with a toddler in the bath, make sure that you do not allow any soap-suds near the eyes and very gently scoop water over the bird, cleaning any particularly soiled places (taking particular care not to miss the fluffy parts around and below the vent). Once satisfied that all is as clean as possible, remove the bird and place it in a second bowl of clean water for rinsing – believe it or not, those birds going to a show on a regular basis soon get used to the procedure and actually appear to enjoy their spa treatment. Make sure that wherever the bird has to stand is not slippery, placing a towel on a draining board, for instance, will help.

Dry the bird by gently patting its feathers with a towel. In days gone by, I used to leave mine in a carrying box overnight in a warm kitchen, but others advocate the use of a hair-dryer set low – if the bird is nervous whilst all this is going on, and you have a penning shed or somewhere similar, it might be worth putting it in one of these and giving an all-over heat from a hair-dryer. Alternatively, if you have an infra-red type of brooding heater, you could set this high above the cage and this will do the same job. As F. E. Wilson has been quoted as saying elsewhere in this chapter, show stock should always be washed a couple of days before the date as, any later than this, their feathers will not have had time to 'web-out' or otherwise restructure themselves.

Cleaning feet and nails

As mentioned previously, hard-feathered breeds do not normally have a bath prior to a show. They will, however, most likely require their legs and feet to be washed and gently scrubbed with a soft nail-brush or old toothbrush to remove all residual dirt from underneath the scales. Old hands advise removing ingrained dirt from under the scales with the point of a cocktail stick, with which you can also flick off any dead scales – but please make sure that they are old and dead as it would obviously be cruel and painful to attempt to remove any that are not. Finish off with a thin coating of well rubbed-in petroleum jelly.

Preparing combs and wattles

Again, as when giving birds a complete bath, take great care to avoid getting any water to which soap has been added into the poor creature's eyes. Sponge carefully to remove dirt and then wipe dry and clean with a square or two of absorbent kitchen roll. While some treatments would be classed as 'faking', it is permissible to use baby powder on white lobes and to oil the combs and wattles with baby oil. As you get to know some of the experienced exhibitors, they may tell you a few of their own concoctions which they use to enhance the colour and appearance of the comb – that's the great thing about showing, even though they might be in competition against you, it seems to be in every chicken-keeper's nature to go out of their way to help anyone interested in the fancy.

ENTRY FORMS AND SCHEDULES

I have a show schedule in front of me that has three hundred and five classifications. These include the breed classes for 'Large Fowl Soft Feather', 'Large Fowl Hard Feather', 'Exhibition Trios', 'Utility', 'Novice Classes', 'Juvenile Classes', 'Junior Handling', 'Rare Breeds', 'Soft Feather Bantam Heavy Breed', Soft Feather Bantam Light Breed', 'True Bantams', Hard Feather Bantams', 'Ladies' Classes', 'Egg Classes', 'Selling Pen' (sold by auction), and also various 'Memorial Classes' and 'Challenge Classes for Birds Entered in Other Classes' – all of which can be rather confusing to both novice and expert alike. If you know the classification of your own chosen breed (and the official breed standards will obviously help), it should be a relatively simple matter to complete a show entry form correctly. The forms are generally available either to download from a website or by application to the club or show secretary. The schedule (plus entry form) is usually in the public domain several weeks before the show and there is generally a closing date for entries at the larger exhibitions – although it is just possible that some of the smaller poultry/fur and feather societies will permit entries on the day.

The 'small print'

As well as describing the various classes, the schedule will include other important factors such as rules, conditions and show timings, together with the cost per entry and (should you be lucky enough to win!) the prize money awarded in each section. Each show will have its own rules and condition, but, typically, they will be based on those laid down by the Poultry Club of Great Britain and the schedule might outline various pointers such as 'birds are only to be fed and watered after judging' and that 'exhibitors are responsible for feeding and watering their own birds'. Importantly, it could also state that '…any diseased, ill or infested bird will be disqualified [and] all associated veterinary costs will be met by the exhibitor'. A timetable of the day's intended proceedings should tell

About 'selling pens'

Selling pens and auctions are, as has been shown in Chapter 5, an excellent and safe way of buying good quality stock. Once you have started showing, they can be the perfect means of moving on some of your stock which may, for one reason or another, have become surplus to requirements. Generally, only exhibitors at a show will be allowed to enter a selling pen – and sometimes only those showing more than a certain number of birds in the competitive classes will be allowed to include their stock. 'Lots' containing only a single male bird are rarely allowed, and each 'lot' may only include 'pairs or female only birds' or breeding trios. While there will obviously be an entry fee for each lot, there are usually no premiums, commissions or sale charges applied. A further point to bear in mind is that boxes for transporting the birds must be provided by the seller – so don't take birds intended for sale in your best and most expensive carrying box!

It should be noted that 'selling pens' are different from 'selling classes', in which birds are judged in the normal way and the price of purchase is set by the owner rather than by an auctioneer and public demand.

you when the venue is open for the penning and settling in of your birds, the time that judging starts, a time when any selling classes/auction and prize-giving are to take place and, finally, the time when you are allowed to begin boxing up your livestock in readiness for the journey home.

CARRYING BOXES

Sad youngster that I was, at a time when most boys wanted the latest gimmick that the toy manufacturers of the era could produce, I wanted a terrier and a poultry-carrying basket! The terrier I got despite parental opposition (actually they didn't know of its existence for months because I kept it at a friend's farm) and the basket eventually came via the kindness of my aunt who had one made for me at the local workshop of the RNIB. Prior to its arrival, I had been much embarrassed by the fact that, whilst everyone else turned up at shows with their exhibits in beautiful wicker baskets, mine were transported either in cardboard boxes or in a poorly constructed hardboard container knocked together at home. With the benefit of adult hindsight, I now realise that it matters little what the carrying boxes are constructed of – the most important thing is that they should be large enough to allow the birds sufficient room to turn round and to keep cool. Large boxes will also help in preventing feather damage. Baskets made for the job have the advantage of being solid, but cardboard boxes do not need cleaning and take up no valuable storage space because they can be disposed of immediately after use.

Carrying boxes and crates come in all shapes and sizes

Should you wish, as I did, to own a 'proper' carrying basket or box, there are some extremely

good ones on the market – some are made of aluminium-type material which looks more suited to carrying the instruments of a rock band than a few chickens. They are, nevertheless, perfect for the job and, depending on their size, may be fitted with dividers that allow up to four birds to be carried in individual compartments. Others may be made of plywood or plastic and, although they are less common because they are difficult to clean, it is still possible to buy the wicker basket type.

Whatever your choice, ensure that there is a good base of dust-free shavings in the bottom of the box as this will help in preventing the bird from sliding around during transit; it will also absorb any faeces and reduce the likelihood of your potential prize-winning chicken arriving at the show venue somewhat soiled. A quite revolutionary idea I heard of recently is to get hold of a piece of 'astro-turf', cut it to the size of the box and place it underneath a layer of shavings – this will undoubtedly allow extra grip and can be washed down after use. You see, chicken-keeping might be a traditional hobby but there are always new adaptations to be made and lessons to be learned!

Those mysterious judges!

Who are the judges? And what right have they to decide whether your birds receive a red card for first position, a blue for second, a yellow for third, or a green for 'reserve'? Not just anyone can judge – they are all, without exception, experienced poultry breeders and handlers, and, in addition, have to qualify as a judge by passing various tests (both practical and theoretical) set out by the Poultry Club of Great Britain. It is a time-consuming and dedicated process to become a 'Panel A' judge and thereby be entitled to pontificate on the merits of virtually every breed – to have done so is likely to have taken at least five years. You can, therefore, be pretty certain that, whenever you enter a show, your chickens and bantams are coming under the scrutiny of people who know what they are talking about!

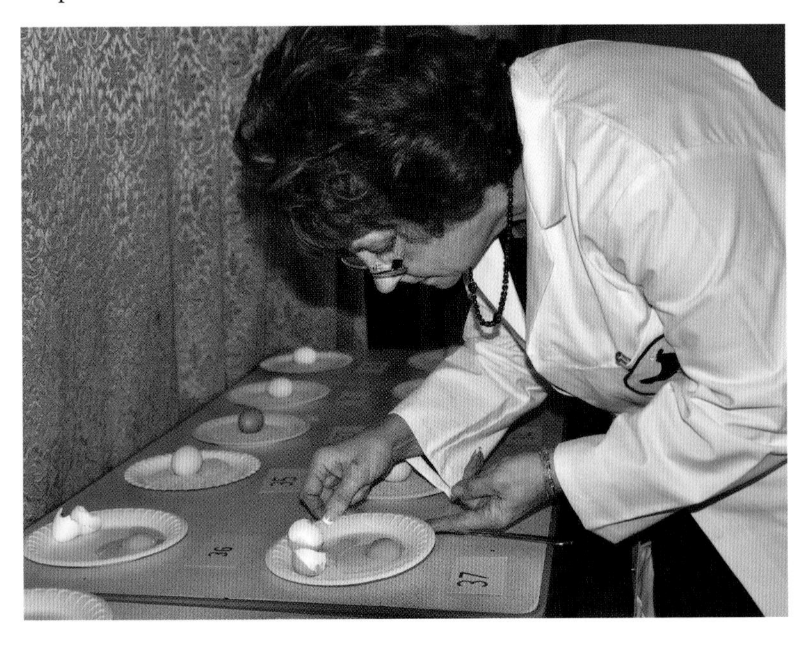

Judging the egg class

EGG SHOWS

Should any of the above prove too daunting – although I sincerely hope that it will have inspired rather than frightened you – there is always the possibility of exhibiting the eggs of your birds. Many clubs and poultry societies will hold an informal egg show at one of their monthly meetings, but there is also the opportunity to exhibit your eggs at most if not all of the county shows that hold poultry classes and, of course, at the large national events.

▼ The egg classes at some shows are very well subscribed

Not just any old egg will do. There are standards attached to eggs just as there are for the showing of pure-bred chickens, and the judges will have to adhere to these when making their eventual selection and giving out awards. You might think that an egg is an egg – but it is not, and, in most cases, it is the shape of it which carries the most points eligible to be awarded by the judges. An egg's shell is almost as important: its texture must be blemish-free and its colour as near to the standard as possible. Its size and weight is also a consideration, as is its cleanliness and freshness. An experienced judge will test for freshness by holding the egg close to their ear and giving a little flick of the wrist in order to hear the sound the contents make. As for having clean eggs, for preference they should not be washed (certainly not with washing-up liquid and hot water!) and many of the experienced egg exhibitors will instead suggest that you remove any marks by rubbing them gently with lemon juice.

Some classes allow for a plate of eggs rather than for individuals, and to have any hope of winning it is necessary that all of them match as closely as possible in shape, colour and size. It is also permissible to enter an egg for its contents, and there are classes in most schedules that stipulate either 'Contents Large Fowl' or 'Contents Bantam' – there are even some marked 'Internal + External Large Fowl' and ditto for bantam breeds. All of this is pretty self-explanatory, as the egg will be judged

A very imaginatively decorated egg! Designed and created by Keith Burgess as a fund-raiser for the Midland Air Ambulance Service

first on its external appearance and then be broken open in order to evaluate the freshness and structure of its yolk and albumen.

As if all that were not enough, you might find that the schedule includes the following: 'Decorated/Painted Egg (Adult)', 'Decorated/Painted Egg (Juvenile)' and 'Decorated/Painted Egg (Junior)'. Sometimes it's possible to let your imagination run wild; at other times the organisers might include a particular theme for egg decoration. The decorated egg classes often contain works of art that would impress even the renowned Russian jeweller Carl Fabergé!

CHICKEN RACING

Of course, rather than exhibit your chickens and their eggs in the conventional way, you might fancy boxing them up and taking them racing in the Derbyshire Peaks! The Bonsall Hen Race (which normally takes place on the first Saturday of August) began almost thirty years ago when the landlord of the local public house decided to change the format of the more traditional hen races previously held throughout Derbyshire (which had been based on the first hen to lay an egg rather than to cross a finishing line).

Anyone with any type of hen can take part in the competition and previously entered breeds range from the common or garden Rhode Island Red, to, appropriately enough bearing in mind the locality, the quite rare Derbyshire Redcap. There can be anywhere between twenty and thirty hens taking part and although most are local, occasionally entrants have come from as far away as Chicago and Gibraltar, the latter having borrowed birds from the pub in order to compete. Race trainers have until 1.30p.m. on the day to enter their birds; racing begins at 2.00p.m. and the organisers aim to finish the championships by about 4.00p.m., after which, the Kimberley Classic Trophy (a large wooden running chicken) is presented to the winning owner. There are also runners up and other prizes given for different elements – such as deportment, appearance and colour.

Some of the trophies, shields and rosettes available to be won at one of the larger shows

As befitting the only organised and regulated race of its type in the world (or so the organisers have it), the rules are strict: birds must begin the race with both feet on the ground and have no assistance from the handler (such as attempting to gain a few precious yards by means of an airborne launch). Any inducement in the way of laying a trail of food along the twenty-metre track is also frowned upon and could well result in immediate disqualification. Horse and greyhound racing it isn't, and the event can be a very slow affair, in which case the race is terminated after about three minutes and the prize given to the hen that is the nearest to the finishing post!

9 FOWL FACTS AND FANCIES

Statistics seem to indicate that the average modern-day Barnevelder hen will lay 200–240 eggs in a season, but Dutch poultry expert Hans L. Schippers tells an interesting story: the inhabitants of Barneveld always claimed that their chickens laid 313 eggs per year. Being a very religious community, no work was done on Sundays – not even by laying hens; 365 days minus 52 Sundays equals 313 eggs!

Most people like to learn of little snippets (or, as Alan Bennett in his play *The History Boys* had one of his characters describe them, 'gobbets') of useless, quirky, but nevertheless fascinating information. Chicken nuggets (sorry, I just couldn't resist!) might include the fact that the greatest number of yolks ever found in one egg is nine; the number of chickens kept in the UK far exceeds the country's human population; or even that an eggshell can have as many as seventeen thousand pores over its surface. I therefore gather here some bits and pieces which, while they might not help you to have any more success with chickens, will, I hope, add to your knowledge and enjoyment of them.

CHICKEN ODDITIES

Chickens and bantams are generally of a certain type; either they are fluffy and cuddly or, as in the case of Old English Game, for example, mean, moody and magnificent. There are, however, others that are known for their unique features: biggest, smallest, long-tailed, no tail at all, amazing crowing abilities or unusual feathering. In an effort to draw viewers, if a chicken-orientated programme were ever to be considered, its producers might very well decide to give their series the title 'Extreme Poultry'. *The Oxford Dictionary* actually describes the word 'extreme' as being '...of a high, or the highest degree...severe...utmost...either of two things as remote or as different as possible'. The same tome points out that the word oddity is a 'strange thing', or a 'peculiar trait'. Some breeds of bantams and chickens are therefore, 'oddities'.

Old English Game: 'mean, moody and magnificent'

Largest and smallest

Although I wouldn't like to state categorically which is the largest breed of chicken, the American-bred Jersey Giant must come somewhere into the equation, as must the feather-legged Brahma, the real Croad Langshan and the many varieties of Orpington (*see* Chapter 4 to find out more about these particular breeds). As for the smallest, there can be no doubt that it is the Serama. Perhaps more correctly known as the Malaysian Serama, the breed was developed in the Malaysian state of Kelantan and they range from between 15 and 25cm (6–10in), and between 225 and 340gms (8–12oz).

▲ The Serama is possibly the smallest of all breeds

Long tail or no tail?

Undoubtedly the prize for the chicken with the longest tail has to be awarded to the Yokohama. In a mature bird, the tail can easily reach 60cm (2ft) in length, and in Japan (where the breed originated) fanciers keep the birds in conditions that prevent them from moulting, to encourage the tail to grow as much as 1m (3ft) each year. As to having no tail, several breeds could qualify. Some breeds have no tail at all: the rumpless variety of the Araucana is more akin to the original breed first kept many years ago by the Arauca Indians of Chile (*see* page 75), but today it is the slightly rarer of the two. Also seen only occasionally is the rumples type of Old English Game, but they do exist – so, should you ever see one at a show, it is as well to remember that they are probably not just a poor specimen caught in the middle of the moult!

▼ The Yokohama is known for its incredible length of tail

Wrong way round feathers, or no feathers?

The Frizzle breed (*see* Chapter 4) just has to be the ultimate example of a bird that got up in a hurry one morning and, as a result, put its clothes on the wrong way round! It gets its name from its strange feathering, which should curl towards the head rather than away from it. With fewer, rather than more, feathers, comes the Transylvanian Naked Neck

– and no, its neck is not thus exposed to make life easier for the blood-sucking Count Dracula who is also purported to originate from the same region! The scientific reason is that any reduction in feather mass improves heat dissipation through the naked area, improving tolerance to heat. It cannot, however, by any stretch of the imagination, be called an attractive bird!

Noiseless or noisy?

Very few cockerels of any breed crow more loudly or quietly than others but, for some reason, the Twente, a breed that originated in Holland, has a reputation for inordinately loud crowing. At the other end of the scale it is generally reckoned that Brahma cock birds crow the least. But it is all relative: the shrieking call of a Game bantam cockerel, for example, is far more likely to disturb your Sunday morning sleep than is the dulcet crow of a heavy large fowl species.

Extremely rare, but originally bred for its crowing ability, is the Yurlov Crower. It was used in the crowing competitions which were extremely popular throughout Russia at the beginning of the twentieth century. Towards the end of the century, its fortunes had changed and in the mid 1980s there were as few as two hundred specimens of pure-bred Yurlov chickens registered throughout Russia. Fortunately, the population now numbers thousands rather than hundreds, but it is probably only due to an interest in the breed by a few German poultry fanciers that the Yurlov Crower has come to Europe – and this only in the last two to three decades.

Frizzles look as if they've dressed in a hurry and put their feathers on back to front!

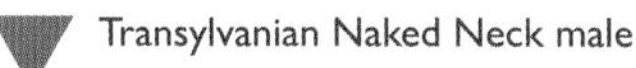

Transylvanian Naked Neck male

FACTS AND FALLACIES

Both abound and sometimes, in the manner of an urban myth, fowl fallacies have a habit of becoming fowl facts if retold often enough! The Internet is sometimes no friend to the chicken-keeper as it helps to perpetuate erroneous information – some of which is potentially harmful, some merely misleading. For example, I have regularly read that breeds with darker coloured feathers and red earlobes will automatically produce brown eggs. Conversely, white hens with white or light feathers and white earlobes are supposed to invariably lay white, or lightly tinted, eggs. It is not necessarily so – whilst it is typical that white-lobed birds will easily produce white eggs, ultimately the egg colour is all down to the genetics of a particular breed.

Pale-coloured eggs do not always come from pale-coloured chickens – nor do brown ones necessarily come from brown hens!

A cock bird will, if he detects potential danger, emit a warning call to his harem which will then either look upwards for the source of danger (a bird of prey, for example) or run for cover. Research has shown that such actions on the part of the male are not involuntary and he will, if on his own, often choose to keep quiet in order to avoid detection. For that reason, it is often thought that a male is a vital addition to your flock; advantageous though it might be, it is not absolutely essential and can sometimes even be detrimental if he spends most of his time pestering and trying to mate with his ladies. If a cock bird is kept, ensure that his spurs do not grow overlong as he may damage his hens while trying to mate. Another commonly held fallacy suggests that spur trimming affects fertility; again, it most certainly does not. Also somewhere near the top of the list of commonly held misbeliefs must be the one which states that a cock bird is necessary to make hens lay eggs – it most certainly is not.

Still on the subject of cock birds, there is an old adage that 'the male is half of the flock'. This is certainly true and he needs to come from a productive and healthy strain which has been proven to achieve a certain objective (as an egg producer, table bird or one of a standard suitable for exhibition). What may be open to some doubt as far as the selection of a male is concerned, is the old idea of picking up a cock and dropping him. Old poultry-keepers used to set great store by this test – a good specimen would apparently stand firm (and even crow) upon hitting the ground, whilst a less worthy potential father would submit to being handled and then run off squawking once on the ground!

When I used to rear literally thousands of game birds annually, I lost track of the number of times that people asked how I kept the eggs warm until they were placed in the incubator and the hatching process started. It is neither necessary nor desirable to keep fertile eggs warm. Indeed, a cool storage place of around 12.5°C (55°F) is ideal. While they are being stored, it will pay to turn the eggs once a day in order to prevent the contents from 'sticking' to the membrane. Most large fowl

Although he undoubtedly adds to the overall attraction of your flock, a cock bird is not necessary if you just want your hens to lay eggs for the breakfast table *(Photo courtesy of David Bland)*

chicken breeds take 21 days to hatch once the incubation period has begun. It is generally accepted that bantams take 19 days; however, some may take 20 to 21. There could be several reasons for this, the most likely being that the eggs were not as fresh as they should have been when incubation was started. Another might be that a large fowl cock was put to a bantam hen (or a bantam cock to a large hen).

It is often said that something is 'as rare as hen's teeth' – due to the fact that chickens do not have teeth and rely instead on the grinding action of their gizzard to masticate their food before it enters the stomach. Chicks, however, do possess an egg tooth (on the tip of the upper mandible) which they use to chip their way out of the egg. This tiny appendage disappears within a couple of days of hatching.

Egg-eating can often occur as a result of an egg becoming broken in the nest box and the resultant mess being pecked at out of curiosity. For some reason, there is a perpetuated myth among the less knowledgeable of chicken-keepers that suggests that it is okay to recycle cleaned and dried eggshells back to their birds, in the fond belief that doing so will provide them with ingredients required to provide strong eggshells in the future. This actually has the potential to do more harm than good. F. E. Wilson, writing a century ago, had this to say: 'Many people do this thinking they are supplying their hens with shell-making material. In not a few instances does the practice of throwing down empty and broken egg-shells to fowls lead to the vice of egg-eating: and not infrequently, as an added vexation, it is the best layers who acquire this bad habit.'

EGGS AT EASTER

Eggs are easily recognised symbols of Easter but their representations of spring, new life, resurrection and fertility were known to our ancestors long before the arrival of Christianity. Once religion took over, eggs – often decorated with symbols such as crosses or fish (the symbol of the early church) – were sometimes consecrated and used in church ceremonies.

▲ Some decorated eggs are far too fancy to be rolled down a hill at Easter!

In some parts of north-east England, eggs decorated for Easter celebrations were known as Pace-eggs (or Paste-eggs), around which many traditions arose. Pace-egg Day is generally accepted as being Easter Monday (Easter Monday was known as Troll-egg Monday in the neighbourhood of Pickering, North Yorkshire) and it was the custom to colour hard-boiled eggs and roll them down hillsides. Among country folk generally, the five latter Sundays of Lent and Easter Day itself were called respectively by the names *Tid, Mid, Miseray, Caning, Palm* and *Paste-egg Day.*

Sometimes pace-egging would take the form of groups of men wandering around the locality, one of whom would have blackened his face with soot and be carrying a basket – the idea being that he and his happy band of followers would persuade villagers to throw boiled eggs into it. When the pace-eggers had received sufficient quantities, they would stop and perform either a short play or dance. These stops were usually conveniently planned so as to occur outside the local inn or the home of a person known to be generous in their hospitality! Meeting a rival band could lead to a fair amount of banter, during which one group often attempted to steal the other's basket of eggs.

Clapping for eggs

During the week before Easter, children in the Anglesey region of North Wales would also go around their village, but, rather than perform a play or dance, they instead begged or 'clapped' for eggs (the custom was known in Welsh as *Clepian Wyau*). They often carried wooden rattles and chanted a little rhyme as they went. Once collected, the eggs were displayed proudly on the dresser at home, with the eggs belonging to the eldest child being placed on the top shelf, those of the second on the next one down and so on. Despite facing strong opposition by late-nineteenth-century teachers who deplored the custom's detrimental impact on school attendance, 'clapping' for eggs remained popular and so, by the early twentieth century, they resigned themselves to the fact that few, if any, pupils would be present in the days before Easter and declared the period to be an official school holiday!

CHICKEN BOOKS AND PRINTS

One way of seeing what the old breeds looked like years ago is to peruse the books, photos and prints that our great grandparents would have had on their shelves or hanging from their walls. Victorian and early twentieth-century poultry-keeping books contain much information that is still of value to today's chicken-keepers and for this reason I have included a few quotes from some of them throughout *Success with Chickens.* Such books are also sought after by collectors and chicken-keepers, as a great part of the pleasure of chicken-keeping is to find old books that are associated with the hobby. Unfortunately, in good condition, they can be quite expensive (especially first editions), but as well as second-hand booksellers' lists and websites, it is often possible to pick up such books in the most unexpected of places. As an example, as a teenager, I used to borrow *Bantams and Miniature Fowl* by W. H. Silk regularly from our local library. Imagine my pleasure and excitement when, more than forty years later, I happened to walk into a rather rundown-looking charity shop in Scotland and discovered an extremely well-preserved first edition, which I bought for seven pounds. Some dealers have it on their website for fifty pounds plus, and it is now an important part of my already extensive collection.

An assortment of old poultry books – from which there is always much of value to be gleaned *(Author's photo)*

Photographs

Poultry show reports have, for many years, been accompanied by photographs and, as such, provide a useful reference for breed standards. Professional poultry photographers are many, but good ones are few and far between. The skill of a photographer who understands poultry and knows just when to press the button is undeniable, but any of them will, I am sure, admit to the fact that digital cameras have made life considerably easier. Prior to such times, the best a professional could do was to set up his subject, use a Polaroid to gain an instant idea of layout and then shoot off reams and reams of expensive film in the hope that one exposure might produce the perfect shot. Before that, of course, it was all down to patience as the photographer put his subject in place, disappeared under the black cloth and, quite literally, waited to see what would develop.

Prints

To have any idea of how chickens looked before the advent of photography, it is necessary to rely on etchings and prints. As well as general farmyard scenes featuring poultry of all descriptions, there are also examples of cock-fighting prints – many of which were mass-produced as a result of a single bird having achieved a certain notoriety. Wealthy 'cockers' would want a record of this fact and place these prints on their walls – in much the same way as today's teenagers might have a poster of a sporting or singing star adorning their bedrooms. These prints still survive and are worthy of purchase because they are, if nothing else, examples of social history. The most commonly available are of the type known as 'aquatints', the production of which involved the use of a resin dissolved in spirit. This was poured over the surface of a warmed etching plate (generally made of copper), onto which had been applied a layer of wax and into which the etcher had drawn the picture's outline with the point of a sharp needle. As the spirit evaporated, it left tiny granules of resin on the surface. After being printed on special paper, the picture would be hand-finished with watercolours.

J. W. Ludlow was a Victorian illustrator and painter of considerable note and there was a fair amount of excitement among the fancy fowl fraternity when, in the spring of 2011, a collection of Ludlow's original poultry paintings was put into a West Country auction room by his great grandson, Oliver Mousley. As far as anyone was aware, there had only ever been one of Ludlow's original paintings on the market before and so it understandably caused much interest when twenty of them appeared at auction together. Most famously, J. W. Ludlow was illustrator for the great Victorian poultry expert Lewis Wright (of whom more below) and his work appears in Wright's book, *The Illustrated Book of Poultry* (also known as *Cassell's Book of Poultry)* – still considered to be one of the most important poultry reference books of all time. Ludlow's chromolithographs have, over the years, been removed from copies of Lewis Wright's book and have even been reprinted, but it would be a rare thing to find any originals – unless, of course, you know differently!

A selection of old prints from Holland *(Author's photo)*

THE VICTORIAN AND EDWARDIAN INFLUENCE

Despite current popularity, the success and proliferation of many of today's breeds, as well as the day-to-day general practice involved with their keep, owes much to those who have gone before. The Victorians and Edwardians were probably just as enthusiastic chicken fanciers as we appear to be. Like us, hobbyists of the time appreciated bantams in particular, for their utilitarian properties as well as the fact that they required very little space and could, therefore, be kept in small gardens. They were also very much in vogue with the aristocracy, who loved their fine plumage and petite appearance. Generally, the Victorians were immensely fond of exhibiting their bantams and chickens, particularly after cock-fighting was made illegal in 1849, when it seems that many of those who had previously been involved in such activities changed to showing their birds instead.

Some of the wealthiest people employed professional poultry-keepers to take charge of their birds, giving them a cottage and providing them with a uniform of sorts – perhaps a legacy from the days when the outside staff of many 'big houses' would include a cock-fighting trainer resplendent in the family livery. Others, however, were happy enough to see to their own stock. Harrison Weir, Lewis Wright and William Cook were considered to be some of the greatest aficionados of Victorian

The Victorians and Edwardians began to favour some of the lighter breeds originating from the Mediterranean in preference to the traditional heavy utility sorts

poultry – Cook in particular being famous as the originator of the Orpington breed. Queen Victoria herself was not immune to the fascination of poultry and is credited by some with the development of the Cochin breed from birds imported from China.

Given the fact that travel was obviously more difficult in Victorian times than it is now, the dedication of would-be fanciers is quite amazing. Probably the only reason that the still rare Sultan breed ever found its way from its country of origin is because of a Mrs Wells of Hampstead who imported examples of the type from a friend in Constantinople way back in 1854. A decade earlier, Mediterranean breeds that are nowadays common in the UK were brought to the poultry yards of England by Victorian breeders who wished for a light, egg-laying breed to replace, or at least run alongside, heavier, dual-purpose and generally less productive native British breeds.

Chicken sheds and feeding practice

Among the working community, chickens had always been kept as egg and meat producers and pecked a living as best they could in the barns and farmyards of rural dwellings. However, the Victorian upper-class interest in poultry as a hobby and status symbol, coupled with their love of artistic and grandiose construction, meant that the most fortunate of birds were penned in palatial houses situated in a prominent part of the courtyard, or a superbly crafted lean-to built along the side of a walled garden – a location that every visitor would be encouraged to visit. As a result, towards the end of the period, even working-class chicken-keepers began to keep their stock in specially made houses, wired-perimeter runs and to feed them proper rations. This had the result that laying hens began to produce more eggs over a longer season and birds destined for the table put on weight far more efficiently than had previously been possible.

Feeding chickens and bantams became a craft and an art form and, in much the same way as they put a bloom on their favourite hunter or racehorse, the Victorian and Edwardian gentry added all manner of secret

From old feeding ideas come modern-day nutritionally produced pellets

potions to the troughs of their poultry – some based on fancy and some on fact. Boiled linseed will indeed put a gloss on a bird's plumage but of some of the other concoctions then in vogue, the less said the better! Because their birds were now confined in proper houses and runs, the working-class poultry-keeper also needed to devise a feeding programme that would compensate for a chicken's scavenging tendencies. 'Fold' systems that could be moved daily across stubble fields in the autumn were all well and good, but the occupants still needed a supplementary diet at other periods of the year. Hot mashes of barley, oats and household scraps were devised – the components of which were then developed through the early twentieth century into the concentrated foodstuffs we give our poultry today.

MODERN-DAY LEGISLATION

To bring things right up to date and in order to head towards a conclusion (of sorts!), it is perhaps necessary to mention certain legislation that may have relevance to the back garden chicken-keeper. There are certainly some long-held beliefs regarding the legalities of keeping birds in close proximity to urban dwellings but, as has been shown in Chapter 3 (page 36), provided that you are not contravening any local by-laws or going against the wishes of any landlord, it should be as easy to keep birds now as it was in Victorian and Edwardian times. Chickens are not dangerous dogs and, provided you keep them out of the neighbour's garden and from running off down the road causing accidents, there is very little to worry about!

As far as feeding kitchen scraps is concerned, you may not legally do so! The reason is that disease can be spread by such feeding. The main concerns are that meat products might be included in kitchen scraps and, technically, whilst it is perfectly permissible to throw your chickens the outer leaves of lettuce, cabbages and all other greenstuffs as you take vegetable produce from the garden to the kitchen, you are not supposed to feed them once they have been in the kitchen and could possibly have become 'contaminated' by contact with chopping boards and knives used to cut both meat and vegetables. For this reason, you should not allow your free-range birds access to the compost heap (although I would never recommend including meat in the compost anyway as it is a sure-fire way to encourage vermin such as rats).

Should your hobby get a little out of hand and you end up keeping more than fifty birds, you will need to let your local Defra office know, as, in order to maintain some idea of the location of poultry flocks in case of an outbreak of notifiable diseases such as avian influenza, they hold a register of all such establishments. With such numbers, you would also need to register as an egg producer, have your premises inspected, and your eggs graded were you to sell your eggs commercially. As a back garden chicken-keeper, there is, however, no such requirement if you just want to sell a few surplus eggs to neighbours and friends. By the way, always put just one little downy feather found in the nest box into your box of eggs as it adds a bit of the 'good life' factor when the purchaser opens the box – a friend of mine used to do this with either a feather or tiny piece of clean straw and his customers loved it!

You might, at some stage, have a few surplus birds to sell on and, if they are pure-bred, your most obvious option is to advertise them through the 'classified' section of either a specialist magazine or your local newspaper. Alternatively, you could try entering them in a selling pen at your local poultry club show (*see* Chapter 8). It is probably never going to be practicable to have your own stand or stall at any public event because the organisers will expect you to have substantial liability insurance cover and, even supposing you can find a company willing to provide this, their premiums will be so high that it makes the whole thing totally unviable.

Food for thought!

Although it is outside the remit of the back garden chicken-keeper, it is as well that everyone knows the current state of the controversial commercial egg market. It seems that some of the EU Member Countries have different criteria, and although the British contingency are doing all they can to keep their egg consumers happy, others are seemingly ignoring any recommendations and/or legislation. In effect, this means that, as NFU president Peter Kendall told delegates at the British Lion Egg Products' press event held in 2010, 'UK producers, including Lion scheme producers, have invested significantly in higher welfare standards to meet the demands of the EU Welfare of Laying Hens Directive… producers in some countries have not. What this means is that the egg industry here is at real risk of being undercut by [overseas] farmers who can sell non-compliant, and let's face it illegal, eggs and egg products.' Food for thought indeed!

10
A CONCLUSION – OF SORTS!

Who would have thought that keeping chickens would ever become a lifestyle objective rather than, as was the case a few years ago, a hobby generally indulged in at the bottom of the garden by a few middle-aged men? Can there be any greater surprise (and pleasure!) than the recent interest, and in the most unlikely of places too. The leisure pages of national newspapers and weekend magazines, which are not normally thought of as being particularly poultry orientated, periodically give space to passionate chicken-lovers who have a pair of tiny Barbu d'Uccle bantams sharing their London mews courtyard, or a couple from urban Sheffield extolling the virtues of a weekend course which persuaded them to buy 'Milly', 'Molly' and 'Mandy' – a trio of hens bought to give them a taste of fresh eggs and self-sufficiency.

Even normally staid BBC Radio 4 recently thought the subject worth an amusing light-hearted programme interviewing some people who,

'People who do love chickens, really do love them!' Exhibitor Robert Armes and his prize-winning Orpington large fowl

while they might well be considered sane in normal life, seemed somewhat off the scale when it came to the delights and antics of their pet chickens! One interviewee was obviously, quite literally, a hen-pecked husband and he recounted the story of how his wife started with just four chickens in the garden and, in a very short space of time, had built up her flock to forty-seven. His tone was of resignation; he had an exasperated, 'if you can't beat 'em, join 'em' attitude. Still, as he said, it made the task of choosing his wife's birthday, Christmas and anniversary presents very easy as he knew she'd be happy with a new coop, roll of wire netting with which to extend the run, or indeed anything at all chicken-related! Some who featured on the programme even admitted that they'd given up annual holidays in favour of staying at home with their birds – and those that did take a break would only go to chicken-keeping relatives and friends who had the facilities to house their birds (carried in a cat basket amongst all the more usual holiday luggage). As the presenter remarked, 'people who do love chickens, really do love them!'

In early 2011, a new monthly magazine called *Your Chickens* appeared alongside other well-established poultry-related publications in order to cater exclusively for the ever-growing followers of back garden chicken-keeping. Proof of its success is a comment from its Content Editor, Simon McEwan, who not long after its launch told me that 'Sales of *Your Chickens* in its first few issues have been most encouraging, and we are delighted with the ever-increasing readership. The magazine, being a dedicated publication about back garden henkeeping, clearly fills a gap in the market.'

WHY EVERY HOME SHOULD HAVE SOME!

Hopefully, the reason why urban and suburban chicken-keeping is so popular will have become self-evident when reading through the chapters of *Success with Chickens*. Nevertheless, some of the potential pleasures previously mentioned might be worth a little recap.

Chickens and bantams are extremely adaptable and, with a little forethought, can be accommodated in most gardens. Some people may try and tell you that you cannot have a garden *and* chickens but, provided you're sensible and do not overstock, you most definitely can. Yes, you might have to take a few precautions by preventing access to your vegetable patch or prize-winning flowers (and you might occasionally be frustrated by the fact that your chickens will, on

Looking to the future? *(Photo courtesy of Flyte so Fancy Ltd, Dorset)*

occasions, probably choose to use one of your patio tubs as a makeshift dust-bath), but there are literally thousands of people out there who will tell you that both are possible.

Chickens are easy to keep and, provided you are aware of some of the few pests likely to be encountered such as red mite and the like, and are thus able to deal with a minor problem before it becomes a major one, there should be nothing in your hobby that will cause sleepless nights. Except, of course, the recurring nightmare of all chicken-keepers who will, at several stages of their lives wake up in a cold sweat wondering whether they've remembered to shut their birds in to protect them against the fox!

Parents quite rightly want their children to realise where their food comes from and while the various television programmes hosted by the likes of Jamie Oliver and Hugh Fearnley-Whittingstall (who between them have probably done more than anyone to expose the practices of commercial farming and encourage public awareness) are excellent, there can be no substitute for first-hand experience.

Chickens are, then, both pets and egg providers. After a day at work, can there be anything better than going down to the hen house, feeding the chickens and collecting the day's fresh eggs? I rather think not. On a summer's evening, open the run door, pull up your patio chair and relax with a gin and tonic. Before many seconds have passed, you will be joined by your chickens looking inquisitively upwards and vocally enquiring about your day – it is, however, probably only out of politeness because, as you will soon discover, they have much more to tell you about theirs!

TEN YEARS HENCE

While it is quite right to celebrate the pleasures of chickens and to give them due prominence as a useful family pet (I wonder whether their popularity will ever surpass that of the dog or cat?), in a 'conclusion', one should, I suppose, also offer a word of caution and a forecast for the future. I worry about the possible end result of this current chicken-keeping phenomenon; especially the practice of penning together (*see also* page 52) types of different pure-breeds because of their colour variations and then, somewhere along the line, deciding that it might be quite fun to breed from them all and, as a result, ending up with a multitude of cross-bred offspring. If that's as far as it goes, all well and good, but we all have a responsibility towards the continuation of the pure-breeds, the genetics of which could be lost if birds of mixed varieties continue to be further crossed. It is not such a far-fetched scenario; in fact we've been down that route in the past when, at certain points during the last century, the bloodlines of pure-breeds were occasionally sacrificed in favour of a hybrid which would produce more eggs and/or put more meat on its breast. Between the two World Wars, the (then) Ministry of Agriculture – now Defra – set up a committee to investigate the situation and reported that the biggest single reason why the efficiency of chicken production was in decline was faulty breeding. Too much emphasis was being placed on selecting for egg and table production and not enough on breeding for hardiness, vitality and health. That so many of today's pure-breeds are still in existence is due only to the efforts of a few individual dedicated fanciers who did all they could to ensure that their gene banks were preserved.

Only time will give the answer to one particular worry and that is what might eventually be the effect of unscrupulous or, more likely, inexperienced, breeders who see a quick profit to be made from the current chicken-keeping trend? We need to ensure that all the careful breeding and maintaining of standards by the fanciers of a generation or so ago is not undone by those who buy, hatch and sell anything, no matter how far away it is from an accepted standard, simply in order to supply the demand. Taken to extremes, it is possible that, with so many people jumping on the band-wagon, there could even be a glut of unwanted chickens in the future as demand is eventually sated.

Those new to the idea of chicken-keeping are sometimes persuaded to keep 'rescue' birds that are sold off after a year in battery cages or similar. The buyers often don't realise that they're at risk of importing disease to their sites, and it is no coincidence that organic standards generally look unfavourably on ex-battery hens and the like. For the future then, it is important to ensure that, with certain honourable exceptions such as accredited rescue groups, all stock (be it pure-breeds or free-range hybrids) is bought from reputable breeders who can guarantee the health of their flocks.

Done sensibly, keeping chickens allows almost anyone the opportunity of a lifetime's enjoyment and education. As Dr David Waltner-Toews, veterinarian and professor at the University of Guelph, Ontario, Canada, is quoted as saying: 'if we do not make room for these urban entrepreneurs, we risk losing a set of very important food-rearing skills that will enable us to better navigate the economic, climatic and environmental instability our society will face in the coming decades.' While this might make you feel that he's talking about Armageddon rather than the benefits of a few back garden chickens, he is making a valid point.

As I left Young James with his chickens, it struck me that his love for them will last a lifetime and it was clear that his little flock of treasured hens had been elevated several places higher than those that are compelled to eke out a living in the farmyard or around the gamekeeper's sequestered woodland dwelling. Never will any of James' birds 'scratch about the mead' as Jefferies' writings have it (Richard Jefferies, the Victorian author famous for *The Gamekeeper at Home*; *The Amateur Poacher*, *Bevis* and *Amaryllis at the Fair*) but I wonder what life a fowl will have several years hence; cosseted as they are at Audley Hall or mere providers of eggs and meat living semi-wild in the barn as some say Nature intended. No matter, the happiness of one boy and his hens was plain to see.

Taken from *Visiting Home* by J. E. Marriat-Ferguson, published privately in 1905

IN CONCLUSION

Have I had success with chickens in my life? Well, if success is defined by unlimited pleasure and an undeniable sense of well-being brought about by their care and antics for not far off fifty years, then yes, I have. Would I emphatically encourage anyone with a few suitable square metres of back garden to indulge in this most obsessive of hobbies? Yes, provided that they go into it with adequate knowledge and understanding, I most certainly would!

BIBLIOGRAPHY AND SOURCES

Dickson Wright, Clarissa: *Rifling Through My Drawers*. Hodder & Stoughton (2009)

Eley, Geoffrey: *Home Poultry Keeping*. A & C Black (2002)

Fergusson Blair, N. (Mrs): *The Henwife: Her Own Experiences in Her Own Poultry Yard.* Original publisher unknown (1st edition circa 1800)

Graham, Chris: *Choosing and Keeping Chickens*. Hamlyn (1st Edition 2006)

Hobson, Jeremy & Lewis, Celia: *Choosing and Raising Chickens*. David & Charles (2009)

Hobson, Jeremy & Lewis, Celia: *Keeping Chickens – Getting the Best from Your Chickens*. David & Charles (revised 2nd edition 2010)

Mannering, Rosslyn: *Fowls and How to Keep Them*. Cassell's (1924)

Marriat-Ferguson, J. E.: *Visiting Home*. Published privately (1905)

Maycock, S. A. (Ed.): *The Smallholder's Pocket Guide*. C. Arthur Pearson Ltd (1950)

Popescu, Charlotte: *Chicken Runs and Vegetable Plots*. Cavalier (2009)

Porter, Valerie: *Domestic and Ornamental Fowl*. Pelham (1989)

Raymond, Francine: *Keeping A Few Hens in the Garden*. Kitchen Garden (1998)

Roberts, Victoria: *British Poultry Standards*. Wiley-Blackwell (2008)

Scrivener, David: *Starting with Bantams*. Broad Leys Publishing (2002)

Silk, W.H.: *Bantams and Miniature Fowl*. Poultry World (1951)

Trevisick, Charles: *Keep Your Own Livestock*. Stanley Paul & Co. Ltd (1978)

Verhoef, Esther & Rijs, Aad: *The Complete Encyclopedia of Chickens*. Rebo (2003)

Wilson, F. E.: *Poultry Keeping and How to Make It Pay*. C. Arthur Pearson Ltd (1903)

Wright, Lewis: *The Illustrated Book of Poultry*. Cassell's (circa 1870)

GLOSSARY

Addled A fertile egg, the embryo of which has died during incubation

Air cell Air space found at the broad end of the egg. It can be seen when candling and its size denotes freshness and, during incubation, the development of the embryo

Albumen The white of an egg

AOC Showing term, normally seen on a schedule and an abbreviation for 'Any Other Colour'

AOV 'Any Other Variety'

Ark Small, portable chicken house

Axial feather Short wing feather found between the primary and secondary flight feathers

Baffle boards Boarding placed around the bottom of a run or wired partition to prevent draughts or birds from seeing each other

Barring Plumage markings of equal space and width across a feather

Beard Tuft of throat feathers

Blastodisc Point of fertilisation on egg yolk (sometimes referred to as the 'germinal disc')

Bloodspot Sometimes seen in the laid egg, the reasons for its appearance could be genetic or nutritional causes

Boots Feathers growing down the legs and across the toes

Boule Type of feather formation found on the necks of some continental breeds

Brassica Vegetables such as cabbage, the leaves of which (in small quantities) make for good chicken feeding

Brassiness When the feathers of the neck and back of light-coloured or white birds develop straw-coloured areas. This is caused by sunshine or extreme weather

Breed Chickens related by ancestry, and breeding true to characteristics such as body shape and size

Brooder An artificial heater for rearing young birds

Broody As an adjective, having the natural instinct to sit on eggs once a clutch has been laid. As a noun, a hen bird showing the same inclination, who will sit either on her own eggs or those substituted

Caecum One of two intestinal pouches found at the junction of the small and large intestines

Candling Using a source of light behind the egg to detect freshness, hatchability and flaws in the shell structure before incubating. During incubation, eggs can be candled to check fertility and humidity

Cap The back part of a chicken's skull

Cape Feathers running from the back of the head down to the shoulders

Carriage A show term denoting the ideal stance of a particular breed

Chalazae 'Shock absorbers' of albumen around the yolk of an egg

China eggs Sometimes referred to as 'pot' or 'dummy' eggs. They are used to encourage young birds to lay in the nest boxes and are also placed under a broody hen to settle her in to the sitting coop before replacing them with fertile eggs

'Clears' Infertile eggs found after candling

Cloaca 'Collection' point for the bird's excrement before it is finally evacuated

Cobby Term used to denote a short-backed rounded body

Cock A mature male after its first moult

Cockerel Male bird before it is known as a cock! More specifically, less than twelve months old

Comb Generally, the horny muscle which appears on the heads of most breeds. Specifically, there are several different types:
Cushion almost circular (Silkies)
Horn two spikes appear at the top of the comb (La Fleche)
Leaf (Houdan)
Mulberry same as *'Cushion'*
Pea three small combs lying parallel with the centre one forming the highest point (Brahma)
Raspberry like half a raspberry! (Orloff)
Rose broad, solid comb, nearly flat on top, covered with several small regular points and topped off with a leader or spike (Rosecomb; Sebright)
Shell same as *'Leaf'*
Single two types, *'large'* (Ancona) and *'folded'* (Leghorn)
Strawberry like half a strawberry! (Malay)
Triple same as *'Pea'*
Walnut same as *'Strawberry'*

Coop Small hutch usually used to house a broody hen when sitting and/or her chicks

Crest Tuft of feathers on the heads of some breeds (sometimes known as a 'top-knot' or 'tassel')

Crop Food collection sac at the internal base of the neck where food is 'softened' before being passed through into the gizzard and digestive system

Culling A sympathetic term for killing or otherwise getting rid of unwanted birds

Cushion Feathers over hen's back near tail

Day-old Despite the obvious, it is actually a term for any chick up to two days of age

Dead in shell A fully formed chick that has died before hatching

Debeaking Trimming back the bird's upper mandible to prevent feather or vent pecking. It should not be necessary with contented chickens and, if practised, would obviously prevent birds from being shown

Defra Government department responsible, amongst other things, for imposing legislation appertaining to chicken-keeping, disease prevention, etc.

Double lacing Term to describe a plumage pattern

Dropping board Removable board fixed under the perches to catch excrement

Dual-purpose A breed that is traditionally thought to be one that will provide both eggs and meat for the table

Dubbing It used to be common practice to cut back the comb and wattles close to the head, especially in Old English Game varieties

Duckwing Colour description of several breeds

Dust-bath A place where chickens can scratch and 'bathe' in order to rid themselves of parasites. They will often create one themselves but it is possible to make an artificial dust-bath from a container of clean, dry sand, pure wood ash, or fine dry earth

Ear lobe May be red, white, blue or purple depending upon the breed

Egg tooth A tiny protuberance at the tip of a chick's beak which is used to break the shell at hatching time

Embryo The young chick in the very early stages of development and before hatching

Face Skin around and below the bird's eyes

'Faking-it' A deliberate attempt by an exhibitor to deceive the judges by dying feathers or covering up faults on a bird

Fold unit Portable combined house and run

Force moult Artificially persuading a bird to moult at an unnatural point of its cycle. Sometimes carried out to ensure a bird is in prime condition for a show on a given date but is not, in my opinion, to be recommended

Furnished A fully feathered bird. The expression is most commonly used to describe a cockerel with full tail and sickle feathers

Gizzard Grinding stomach with muscular lining for pulping food

Gullet Sometimes known as the *oesophagus*, the tubular structure leading from the beak to the crop

Gypsy-face The traditional reference to a dark-purple or mulberry-coloured comb and wattles (as can be seen on a Silkie)

Hackles Long, pointed neck feathers. Also found on the saddle, where the feathers are rounded in the hen and pointed in the cock. (The latter are sometimes known as 'hangers' on a cock bird)

Hard feather Terminology to describe a category of feather type – often seen in show schedules

Harden off Gradually weaning a young bird from either its natural mother, a foster broody or an artificial heat source

Headgear Comb, wattles and ear lobes

Hen Female after her first laying season

Hen feathered A cock bird which does not have either sickle feathers in the tail or proper hackle feathers

Horn Beak colour shadings (especially noticeable in the Rhode Island Red)

Inbreeding When members of the same family are bred together through several generations

Isthmus Part of the oviduct where the shell membranes are included during the formation of an egg

Keel Bony ridge of the breastbone

Lacing Strip of a different colour around the edge of the feathers

Leader Backward-pointing spike found at the back of any bird which is classified as having a rose comb

Litter Floor and/or nest box material

Magnum Part of the oviduct from where the white (or albumen) is secreted during the formation of the egg

Mandibles The upper and lower parts of a bird's beak

Marbled Sometimes known as 'mottling', both indicate spotting on the plumage of breeds such as Anconas

Markings The general plumage pattern of a chicken

Membranes There are two membranes in an egg; they separate at the widest end in order to form the air cell

Moons Spangled markings on the plumage

Mottled (*see* 'Marbled')

Moult The period during which a bird sheds its old feathers and re-grows new ones. Generally occurs in the late summer/early autumn

Muff The beard and whiskers found on some breeds

Oil sac When a bird is observed preening itself and continually wiping its beak up the base of the rump, it is sourcing the natural oils found there in the oil sac or gland. The oil sac is essential for feather condition

Ova Yolks in the ovary before falling into the oviduct, from where they become part of the egg

Oviduct The reproductive tube between the ovaries and the cloaca

Pencilling Small marks or stripes on the feathers. Can be straight across, slightly V-shaped, or crescent shaped following the outline of the feather

Pile Colour description of several breeds

Pin feathers In the shooting world, highly prized feather taken from the wing of a woodcock and often used as a paintbrush by artists, but in this context, the term is sometimes used to describe the new feathers emerging after the moult

Primaries The stiff flight feathers found on the outer tip of the wing; there should be ten

Pullet Young hen from hatching to the end of the first season

Quill The hollow stem that connects an individual feather to the body – it is, however, most commonly talked of in connection with the flight and secondary feathers

Recessive Genes which may not be evident in the initial mating, but which may manifest themselves in subsequent generations

Roach back A deformity of the vertebrae showing as a hunched back

Saddle Posterior part of the back of a male bird, equivalent to hen's 'cushion'

Scales Found on legs and toes

Secondaries The wing feathers which can be seen even when the wing is folded

Self-colour Plumage of the same colour throughout

Serrations The indentations noticed on a bird possessing a single comb

Shaft The stem or base of any feather. On newly moulting birds, blood can often be seen internally through the shaft

Shank The leg of a bird

Sickles Long curved feathers on the outer sides of a cock bird's tail

Slipped wing Where the primary feathers hang below the secondaries when the wing is in a natural position

Spangling Plumage in which there is a different colour towards the tip of the feathers

Split wing A deformity that shows itself when the axial feather is missing. It is thought to be hereditary and it is therefore best not to breed from any bird displaying this fault

'Sport' Any bird with naturally occurring colour or

characteristics which differ from the accepted breed standard

Spur The pointed, horny projection at the base and rear of a cock bird's legs. The spurs on a cock grow longer as it matures but this is not, as sometimes thought, a reliable indicator of the bird's age

Strain A group or flock of chickens carefully bred over several generations by an individual fancier. Quite often, the strain will be known by the name of the breeder. Some producers of particular breeds may have separate 'cock breeding' and 'pullet' breeding strains

Stub An immature, partially grown feather

Tail feathers The stiff straight feathers of the cock bird normally found under the sickle feathers. On some breeds, however, notably the Cochins, Brahmas and Orpingtons, they are absent

Trachea Otherwise known as the windpipe, the part of the respiratory system that allows air to pass from the larynx to the lungs and bronchi

Under colour Colour of the downy part of the plumage that is normally only seen by gently brushing back the outer feathers

Uterus The part of the oviduct where all the individual elements of an egg come together during its formation

Utility Like a dual-purpose breed, one which is recognised to be useful as both an egg layer and table bird

Variety Sub-division of breed recognised by its colour or plumage pattern

Vent Rear 'opening' through which droppings and eggs are excreted

Wattles The folds of skin hanging either side of the lower beak. Ideally they should be of equal length

Wing clipping Sometimes it is necessary to clip the primary and secondary feathers of one wing to prevent the lighter breeds from flying. The feathers will re-grow in the next moult but until then it would not be possible to show the birds

Wry tail Any tail that is carried either to the left or right of the imaginary continuation of the backbone is said to be 'wry-tailed'. It can affect both hens and cocks and neither should be bred from, because it is a genetic defect

0019
0020
0021
0022
0023

INDEX